AV INSTRUCTIONAL TECHNOLOGY MANUAL FOR INDEPENDENT STUDY

AV
INSTRUCTIONAL
TECHNOLOGY
MANUAL FOR
INDEPENDENT
STUDY

Sixth Edition

Edited by
JAMES W. BROWN
Professor Emeritus of Instructional Technology
San Jose State University

RICHARD B. LEWIS
Late Professor Emeritus of Instructional Technology
San Jose State University

Prepared by
William G. Allan, H. Lois Brainard,
James W. Brown, James Cabeceiras,
George W. Cochern, Leonard J. Espinosa,
Harold H. Hailer, Fred F. Harcleroad,
Ronald L. Hunt, Leslie H. Janke,
Jerrold E. Kemp, Richard B. Lewis,
Ronald J. McBeath, Richard S. Mitchell,
John E. Morlan, Glen Pensinger, Earl Strohbehn

McGraw-Hill Book Company

New York   St. Louis   San Francisco   Auckland
Bogotá   Hamburg   Johannesburg   London   Madrid
Mexico   Montreal   New Delhi   Panama   Paris
São Paulo   Singapore   Sydney   Tokyo   Toronto

**AV Instructional Technology Manual
for Independent Study**

1 2 3 4 5 6 7 8 9 0   S E M S E M   8 9 8 7 6 5 4 3

ISBN 0-07-008179-4

Library of Congress Cataloging in Publication Data
Main entry under title:

AV instructional technology manual for independent
   study.

   Bibliography: p.
   1. Audio-visual education. 2. Audio-visual materials.
I. Brown, James W. (James Wilson), date
II. Lewis, Richard Byrd, date   III. Allan,
William G. IV. Title: A.V. instructional technology
manual for independent study.
LB1043.A9 1983   371.3'3   82-14868
ISBN 0-07-008179-4

This book was set in Helios by Allen Wayne Communications, Inc.
The editors were Christina Mediate and James R. Belser;
the designer was Robin Hessel;
the production supervisor was Phil Galea.
New drawings were done by J & R Services, Inc.
Semline, Inc., was printer and binder.

THE AUTHORS

All the authors, with exceptions noted below, are (or were, to the time of their retirement) members of the faculty of the School of Education, San Jose State University.

William G. Allan, M.A.
Assistant Professor Emeritus of Education,
Department of Instructional Technology

H. Lois Brainard, Ed.D.
Associate Professor of Education,
Departments of Instructional Technology,
Elementary Education, and Early Childhood Education

James W. Brown, Ph.D.
Professor Emeritus of Instructional Technology

James Cabeceiras, Ph.D.
Professor of Education,
Department of Instructional Technology

George W. Cochern, Ph.D.
Professor of Education,
Department of Instructional Technology

Leonard J. Espinosa, Ph.D.
Professor of Education,
Department of Instructional Technology

Harold H. Hailer, Ph.D.
Professor of Education and Chairman,
Department of Instructional Technology

Fred F. Harcleroad, Ph.D.
Professor of Higher Education,
University of Arizona, and formerly Dean of the College,
San Jose State University

Ronald L. Hunt, Ed.D.
Professor of Education,
Department of Instructional Technology

Leslie H. Janke, M.A., M.L.S.
Professor Emeritus of Education and Library Science,
and Director, Division of Library Science

Jerrold E. Kemp, Ed.D.
Professor of Education,
Department of Instructional Technology,
and Coordinator of Instructional Development,
Instructional Resources Center

Richard B. Lewis, Ed.D.
Late Professor Emeritus of Instructional Technology

Ronald J. McBeath, Ph.D.
Professor of Education,
Department of Instructional Technology, and Director,
Instructional Resources Center

Richard S. Mitchell, Ed.D.
Professor of Education,
Department of Instructional Technology

John E. Morlan, Ed.D.
Professor of Education,
Department of Instructional Technology

Glen Pensinger, M.A.
Television Engineer, Instructional Television Center,
Instructional Resources Center

Earl Strohbehn, Ed.D.
Professor Emeritus of Education,
Department of Instructional Technology

CONTENTS

SECTION TWO: CREATING
INSTRUCTIONAL MATERIALS   43

SECTION THREE: OPERATING AUDIOVISUAL EQUIPMENT   115

SECTION FOUR: CORRELATED REFERENCES   160

SECTION FIVE: PERFORMANCE CHECKSHEETS   167

PREFACE

A.T.&T. Co. Photo Center

This manual is not a textbook. Nor is it a substitute for a text or for other basic references in instructional technology, media, and methods. The manual will be most effective if combined with the direction and contributions of an instructor.

Nearly all the exercises offer opportunities for you to work alone. But some are presented in ways that make possible group study and production activities. All the exercises provide things to do in order to give you experiences in the practical solution of problems of choosing, using, and inventing instructional materials for your own instructional procedures and for your students.

Readings and References

Information basic to completing many of the exercises in this manual is contained in the correlated textbook, *AV Instruction: Technology, Media, and Methods,* 6th ed., by James W. Brown, Richard B. Lewis, and Fred F. Harcleroad, McGraw-Hill, 1983. Other selected readings are recommended in the Correlated References section at the end of this manual.

Learning about Media

Most colleges and universities provide courses in the use of educational media and technology. Instructional procedures for them vary. To implement the many options now available in such courses, this manual is designed to be used in any of the following situations:

- Several courses taught in the same term
- A sequence of courses taught over a period of time
- Courses dealing with general methods of organizing and conducting instruction
- Courses directed to the study of teaching methods for special subject fields

- A concentrated, single basic course in media resources—their selection, production, and utilization
- A correspondence or continuing-education course in audiovisual methods and materials
- A basic course for students in community colleges preparing for media technician responsibilities

Organization of This Manual

Following an introductory section, "A Systematic Approach to Planning with Media," this manual is organized in five sections, each dealing with a specific aspect of the field, as follows:

☐ **Section One:** *Selecting and Using Ready-Made Materials.* Media center characteristics, reference resources, motion pictures, television, audio materials, filmstrips, flat pictures, free and inexpensive materials, textbooks, programmed materials, computer programs, simulations and games, maps and globes, multimedia kits, community resources and field trips, and self-instruction facilities.

☐ **Section Two:** *Creating Instructional Materials.* Copyrights; drawing and sketching; displaying; graphs and time lines; cloth, magnetic, and electric boards; mounting and laminating pictures; duplicating; making transparencies; inventing games and simulations; puppets; editing and making audio tapes; visual literacy; copying with cameras; making still pictures; storyboarding; making motion pictures; television production; media in testing; and creating self-paced modules.

☐ **Section Three:** *Operating Audiovisual Equipment.* Audio disc players, audio tape players and recorders, projection principles, overhead projectors, 2- by 2-inch slide projectors, filmstrip projectors, opaque projectors, 16mm and 8mm motion picture projectors, television receivers, video players and recorders, microcomputers, and equipment maintenance.

☐ **Section Four:** *Correlated References.* This section presents up-to-date references in 42 categories that are correlated with the various exercises of this manual.

☐ **Section Five:** *Performance Checksheets.* The checksheets are designed to guide your practice with audiovisual equipment, to facilitate your learning to operate them on a step-by-step basis, and to provide a record of your progress toward achieving such skills.

We hope that you will become expert in choosing, using, and producing the basic educational media resources treated in this manual and that, as a result, you will enjoy and be more effective in your future teaching or training assignments.

James W. Brown
Richard B. Lewis

AV
INSTRUCTIONAL
TECHNOLOGY
MANUAL FOR
INDEPENDENT
STUDY

INTRODUCTION

A SYSTEMATIC APPROACH TO PLANNING THE USE OF EDUCATIONAL MEDIA

Application of systematic procedures recommended in this section will aid in the instructors' tasks of searching for, selecting, obtaining, producing, supervising production of, and using educational media resources and related equipment.

Characteristics of the Approach

Examine Chart 1, "The Systematic Approach of Instructional Technology." Note that its central focus is upon *students*— their needs, capabilities, and achievements—as they work toward desirable levels of competence or performance. Note, too, that the chart calls for answers to four fundamental questions: (1) What goals are to be achieved? (2) How, and under what conditions, will students seek to achieve those goals? (3) What resources are required for necessary learning experiences? (4) Outcomes: How well were goals achieved? This process also provides guidance in necessary improvements in the instruction: What needs to be changed?

A MODEL FOR SYSTEMATIC PLANNING OF INSTRUCTION

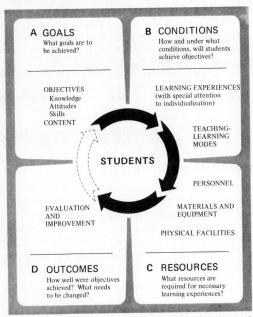

In the chart, all planning is centered on students, their needs, their capabilities, and their achievement of carefully specified objectives. The four quadrants of the chart contain the specific planning steps, each of which is necessary and important. Planning for teaching is one principal role of any person who teaches.

To answer the questions, seven steps in instructional development are recomended:

1 Define (or accept) objectives and select content to be studied.

2 Select appropriate learning experiences and seek to individualize them.

3 Select one or more appropriate teaching-learning modes in which to carry out learning experiences.

4 Assign personnel roles.

5 Select appropriate materials and equipment.

6 Choose physical facilities for the learning experiences.

7 Evaluate results and recommend improvements.

Structuring Learning Experiences

Examine Chart 2, "Relationships of Objectives and Teaching-Learning Modes." It stresses two points: There are different types of teaching and learning objectives; and there are different modes in which to organize and conduct learning experiences to achieve those objectives. The lower section of the chart suggests that there are different kinds of instructional modes—those involving the teacher alone, those in which student-teacher interaction or student-student interaction is present, and those in which *students* work alone. What are the implications of this classification in determining the proper role of media in teaching-learning situations?

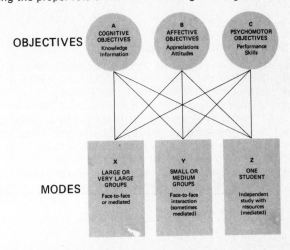

Experiences Leading to Learning

Chart 3, "Experiences Leading to Learning," enumerates the many different types of activities which, if thoughtfully and appropriately planned, can lead to desired learning. Examine the list. Has anything been omitted which you consider to be an important learning experience? Does this list help to remind you that *many different kinds of learning experiences* can be planned for your teaching and that there is no reason why teaching must be conducted in routine, unvarying ways?

Think about the different kinds of classroom teaching you have experienced in your own school career. Does this list remind you that activities of a typical class period will usually involve several interrelated experiences?

Chart 3

EXPERIENCES LEADING TO LEARNING

Thinking
Discussing, conferring, speaking, reporting
Reading (words, pictures, symbols)
Writing, editing, scripting
Listening
Interviewing
Outlining, taking notes
Constructing
Creating
Designing, drawing, painting, lettering
Photographing
Displaying, exhibiting
Graphing, charting, mapping
Demonstrating, showing, presenting, informing, instructing

Chart 3

EXPERIENCES LEADING TO LEARNING (*Continued*)

Experimenting, researching
Problem solving
Collecting
Observing, watching
Traveling
Exchanging
Audio recording
Video recording
Dramatizing
Singing, dancing
Imagining, visualizing
Organizing, summarizing
Computing, programming
Judging, evaluating
Working

Educational Media for Learning

Now study Chart 4, entitled "Educational Media for Learning." Does this chart—together with Chart 3—remind you that learning can best be achieved by a variety of experiences in which a great variety of media are used? Reexamine the list. How many of the items in it are essentially *verbal* in nature? How many involve only elements of *sound*? Which of them are comprised chiefly of representational *visual* elements? *Sound* plus *visual* representations in combination? Elements selected from *real things*? Elements to be used in *making things* to promote learning? How many of the items are likely to be available in ready-made form? How many will be created by teachers or students?

Chart 4

EDUCATIONAL MEDIA FOR LEARNING

Textbooks	Filmstrips
Supplementary books	Microfilms, microcards
Reference books, encyclopedias	Stereographs
Magazines, newspapers	Maps, globes
Documents, clippings	Graphs, charts, diagrams
Duplicated materials	Posters
Programmed materials (self-instruction)	Cartoons
	Puppets
Motion pictures (films, video tapes, discs)	Models, mock-ups
	Collections, specimens
Television programs	Flannel-board materials
Radio programs	Magnetic-board materials
Audio recordings (tapes and discs)	Chalkboard materials
	Construction materials
Computer courseware	Drawing materials
Flat pictures	Display materials
Drawings and paintings	Multi-media kits
Slides and transparencies	

Equipment for Learning

Now examine Chart 5, "Equipment for Learning." Its significance is best grasped by relating the items contained in it to those listed earlier in Charts 3 and 4. Use of media often requires instructional equipment. A film cannot be seen without a projector, and an audio recording cannot be heard without a suitable playback device; a television program requires a receiver and often a video tape player. Can you think of any equipment items that should be added?

Chart 5

EQUIPMENT FOR LEARNING

Audio record players, tape recorders, radios
Slide and filmstrip projectors and viewers
Overhead projectors
Motion picture projectors and viewers
Television receivers
Video-tape recorders, players, viewers, video-disc players and recorders
Teaching machines
Microcomputers, computer terminals, peripherals
Electronic laboratories, audio/video access and interaction devices
Telephones with or without other media accessories
Microimage systems—microfilm, microcard, microfiche
Copying equipment and duplicators
Cameras—still and motion

Media in the Teaching-Learning Sequence

Chart 6, "Media in the Teaching-Learning Sequence," emphasizes that educational media and experiences of many different kinds are used in various ways and for various purposes in five phases of the teaching-learning sequence: (1) introduction, (2) development, (3) organization, (4) summary, and (5) evaluation.

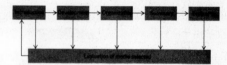

MEDIA SELECTING IN PHASES OF TEACHING

Physical Facilities for Learning

A further element of key importance in the process envisioned for the systematic approach to teaching and learning involves the characteristics and appropriateness of physical facilities in which learning activities are to be conducted. Study the various types of physical environments suggested in Chart 7, "Facilities for Learning." Can you think of other types of facilities that should be listed?

Chart 7

FACILITIES FOR LEARNING

Lecture halls	Theaters
Classrooms	Studios
Divisible	Libraries
Undivided	Resource centers
Independent study areas	Electronic learning centers
Discussion rooms	Playing fields
Laboratories	Community resources
Shops	Home study centers

SECTION ONE

SELECTING AND USING READY-MADE MATERIALS

Those who select and teach with ready-made instructional resources soon recognize that these resources vary greatly in scholarship and truthfulness, in degree of appropriateness for intended audiences and educational purposes, in cost, and in ease or feasibility of application. These and other characteristics of instructional materials—all of which are prepared by others, usually commercial producers or publishers—deserve professional attention. Giving this attention is an important part of the systematic approach of instructional technology.

Media in the Teaching-Learning Sequence

The importance of ready-made media in the teaching-learning sequence is apparent. Typically, media resources are selected and used for different purposes in each of five phases:

☐ **Introduction Phase** (motivational, exploratory). Here, especially, materials are sought which raise questions, excite interest, provide interesting overviews, and allow students to demonstrate present knowledge or skill.

☐ **Development Phase** (goal setting, location and study of informational materials, individual and group work toward solution of problems). Materials especially sought during this phase provide essential data for answering questions as they arise.

☐ **Organization Phase** (pooling of results of research and study; presentation and integration of findings). Materials needed here provide integrated review of studies, fill in gaps in understanding, and prepare for presentations of findings during the summary phase, which follows.

☐ **Summary Phase** (a series of unit culmination activities). The summary phase sometimes involves individual and committee presentations, visual displays, dramatizations, discussions for the purpose of summarizing major ideas, generalizations, or principles developed through the study. Many media resources that are especially useful for this phase are dealt with in Section Two, "Creating Instructional Materials."

☐ **Evaluation Phase** (appraisal of results). Conducted regularly, but especially necessary at the conclusion of a unit of study, the evaluation stage includes testing of students; evaluations of learning products; expressions of student opinions with regard to the continuing value of various activities and approaches used in the study; determinations of changes and improvements needed if a similar study is undertaken with another group; and special attention to uses of *nonverbal* materials in evaluating understandings, appreciations, skills.

General Selection Criteria

Criteria such as the following are often used to guide the selection of educational media:

☐ **Content** Does the item deal with significant curricular content? Is it up to date? Is it accurate? Is it pitched at a proper level of difficulty and sophistication for students with whom it will be used? Is it improperly biased in point of view?

☐ **Purposes** For what significantly important instructional purposes may the item be used?

Clark County School District, Las Vegas.

3

☐ **Appropriateness** Is the medium used suited to the message it seeks to communicate? If the topic is essentially one that requires portrayal of motion, for example, does the medium show motion? Or if color is essential to the message, is the item in color?

☐ **Cost** Is the item likely to be worth what it costs, as measured by educational results derived from its use? Would other less expensive items in other media formats, for example, be the better choice?

☐ **Technical Quality** Is the item technically satisfactory in photography (color, exposure, angles, focal lengths of lenses used), editing, and sound? In type selection, spacing, binding?

☐ **Circumstances of Use** Will the item function effectively in circumstances and environments in which you are likely to use it? Is it suitable for large groups, small groups, individual study? If it must be projected, for example, will images be sufficiently large and bright for all to see?

☐ **Learner Verification** Is evidence supplied that the producer of the item has improved it through systematic trial and revision before offering it to purchasers? Are such data available? Are characteristics of the trial groups sufficiently similar to those with whom it is likely to be used in your situation?

☐ **Validation** Are reliable data available and supplied which prove that students do learn accurately and efficiently through use of the item? Again, are the characteristics of the trial groups sufficiently similar to those of the students with whom the item will be used in your situation?

General Principles

Several general principles apply to the selection and use of media:

- No one medium is best for all purposes.
- Media use should be consistent with objectives.
- Users of media must be thoroughly familiar with them.
- Media must be appropriate for the instructional mode in which they are used.
- One's own preference must not stand in the way of choosing or using particular media.
- Media are neither good nor bad simply because they are concrete or abstract in nature.
- Physical conditions surrounding uses of media may affect significantly the results obtained.

Section One Exercises

Exercises in this section of the manual direct experiences that lead, first, to familiarity with the purposes and services of media centers and with publications that provide information about sources, content, cost, and usefulness of specific resources. Following these two introductory units are seventeen other exercises that offer information about selecting, processing, storing, and teaching with motion pictures (films and video), television, audio materials, filmstrips, flat pictures, reference materials, free, sponsored, or inexpensive materials, textbooks, programmed materials, computer programs, games and simulations, maps and globes, multimedia kits, field trips, and self-instruction facilities and equipment.

EXERCISE 1

MEDIA CENTERS AND LEARNING CENTERS

The Reason for It

Resources of all kinds—both in print and those of audio and visual types—are of fundamental importance to the success of an instructional program. Since trends in education are increasingly directed toward diverse modes of instruction, resources for learning are made accessible by a variety of service organizations and arrangements of facilities. Among these, media centers and learning centers have developed to serve the needs of both teachers and students.

Before You Start

Brown, Lewis, and Harcleroad: *AV Instruction*, Chap. 2. This manual: Exercise 19; Correlated References, Section 28.

Purposes

1 For you to explore the materials and support services provided for teachers and students

2 To permit you to determine the organizations that provide media services: how those services are delivered and obtained

3 To have you learn about learning centers in media centers and libraries: their types and scope

4 To have you investigate learning centers in classrooms: types, purposes, and activities and materials provided

Assignments

1 Plan for and visit one or more of the following media centers, according to your interests and needs:

a A college instructional media or resources center, library, curriculum library or center, instructional resources center, audiovisual center, media production center, instructional television center, photographic services center. (Your selection will depend upon the local situation, and, if required, a conference with your instructor.)

b A school district, county, or regional instructional resources center.

c A single-school comprehensive media center or multimedia library.

d A classroom learning center—or several types of learning centers—in classrooms in one or in several schools.

2 Prepare checklist(s) for your observations; refer to the following suggestions.

3 After your visit(s), organize your findings for a short report to the class or for submission to your instructor. Include in your report such features as: pictures you have taken of service areas and of students at work in learning centers or in reference and study areas; a brief audio tape you have recorded of media center personnel as they explain their services; select excerpts that exemplify the scope and spirit of the services to teachers and students. Prepare a sample col-

San Diego County Schools, California.

Portland Community College, Oregon.

lection of important forms, catalogs, and policy statements of the center(s) for a display to accompany your presentation.

Pointers on Observations

As you make your explorations, watch especially for materials, equipment, and facilities that you believe should be in your classroom or in the school media center. Other items, for reasons such as economy, may be in a school district, county, regional, or other special center. Still others will be in a state or privately operated center such as a motion picture library.

Such general questions as these will guide you in your collection of information: What materials are available for teachers and for students to use? What services are provided for teachers? For students? Where are the materials and services available, and how are they obtained? What are policies and practices in the operation of learning center activities in the media center? In classrooms? More detailed questions to guide your preparation of checksheet(s) appear in the following section.

Develop Your Checklist(s)

In the two groups of questions below on media centers and on learning centers, the italicized headings and suggestions can guide you in developing your own checklist(s):

1 *What is the range of services?* Look for variety, quantity, and condition of materials; methods of cataloging, organizing, displaying materials; provisions for preview; profes-

Project ALOHA.

sional assistance for teachers on selection and utilization.

2 *How are materials selected?* What are the criteria for selection for various types of materials; committees assisting in selection; committee responsibilities and guidelines; budget appropriations for purchases of materials?

3 *Where are materials available?* Find out what is provided each teacher or classroom; what is located in each school library or media center; what is circulated from the *central* instructional-materials center; what is obtained for teachers from sources outside the school or center.

4 *How are materials obtained by teachers?* Determine accessiblity of materials and the time limits imposed upon orders; methods of ordering; how materials are circulated, delivered, picked up; whether trained persons are assigned in each school to help you order materials, use them, and handle mechanical details.

5 *What audiovisual equipment is available?* What equipment is provided in each room, or building, or school; what equipment is circulated by the instructional-materials center; under what conditions are such items loaned to teachers; what assistance is available in connection with equipment?

6 *Are special materials produced?* Find out what is produced and for what purposes, who determines what is pro-

duced, who designs and produces the materials. Are facilities and supplies available for teachers who wish to produce some of their own materials?

7 *What materials selection aids are available for use by teachers?* Do you see evidence of catalogs, bibliographies, study guides, and informative advertising materials? Are there textbooks, manuals, how-to-do-it brochures, and similar items to assist in using instructional materials? Are there current magazines on subject fields and on instructional media and technology and on methods of teaching?

8 *What materials and services are available to students?* At each step of your investigation, determine what, where, and under what conditions students may use materials and services. Determine which personnel of the media center provide assistance to students in their research, project development, and independent study. Are students provided resources to permit them to plan and construct various audiovisual media for use in their own projects?

For learning centers in libraries and comprehensive media centers, and in classrooms:

Dallas County Community College.

1 What subject content and learning objectives are allocated to specialized learning centers?

2 What kinds of activities are provided for each type of learning center?

3 What types of media, materials, and equipment are allocated to each type of learning center? What special facilities are provided, such as special spaces, furniture, and accessories?

4 Are special modular units or packages of resources provided, either by local production units or from commercial sources?

5 Generalize the principal goals for learning centers you observe, in both media centers and in classrooms, in such categories as: general enrichment and recreation, remedial studies and drill, special subject matter centers, film and television activities, production and creative activities.

6 Can you determine whether special efforts and procedures are used to evaluate the results of student use of learning centers?

Charles Beseler Company

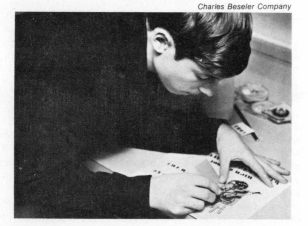

Gage Educational Publishing.

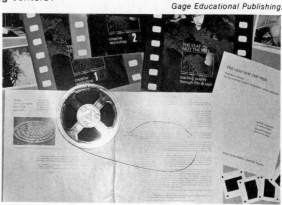

EXERCISE 2

LOCATING INFORMATION ABOUT MEDIA

Control Data Corporation.

The Reason for It

Searching for informational resources need not be a trial-and-accidental-success process. Many useful aids are available to help you find what you need, and if you use them wisely, you will save time and energy. Although finding materials may seem to become increasingly complex as the number of new media resources increases, with the many aids to help you in your search, you may meet this challenge by following suggestions in this exercise.

Before You Start

Brown, Lewis, and Harcleroad: *AV Instruction*, Chap. 17; Reference Section 6. This manual: Correlated References, Section 9.

Purposes

1 To acquaint you with the sources, contents, and value of several well-known indexes, guides, and periodicals used by media specialists in their search for resources that meet their needs and those of their clientele

2 To learn to utilize these references to best advantage in selecting informational resources for use or for purchase

Required Materials

1 A collection of guides, publishers' and producers' catalogs, and compendiums providing *technical* reviews or descriptions of new books, films, filmstrips, and other media resources offered for sale

2 A collection of media-related periodicals, directories, or other publications that contain *qualitative* reviews of new books, films, filmstrips, and other media resources

3 A collection of books, catalogs, reports, or similar publications that provide *qualitative* descriptions and evaluations of various types of media resources considered useful for your needs

Assignments

1 Examine several producers' and publishers' catalogs. Assess the contents of each. How are they organized? What technical information and what descriptive data for separate items does each provide? Were items listed in the catalogs tested and revised prior to release for publication? What test and revision procedures were used, and what data were gathered and reported? Of what value is the information contained in these catalogs in helping you decide whether or not specific items would be suitable for your use?

2 Examine catalogs of several commercial or noncommercial film rental organizations, including one that may be available from the library or media center of your own college or university. How are these catalogs organized? What data are contained in their item annotations? What is stated specifically about qualitative aspects of the items as opposed to merely descriptive or technical characteristics? How would the information included help one determine whether or not to use the items? How, if at all, would it help in determining how to use them? What other data essential to such decisions should be given?

3 Skim the sections of several periodicals, such as *Choice* or *Booklist*, containing evaluative or qualitative reviews of various media. How do their entries differ from those mentioned in Assignments 1 and 2 above? Do they provide enough information to satisfy your needs in ordering items for use in your classes? In purchasing items? Deciding whether or not to request items for preview or examination for possible purchase? What information other than that provided would be desirable or essential for any of these purposes?

4 Complete a similar assignment using one or more book or card directory systems, such as those published by the National Information Center on Educational Media (NICEM), Educators Progress Service, the Educational Film Library Association, Landers Film Reviews, or Audiovisual Associates' *International Index to Multi-Media Information*. How do their entries differ from those mentioned in Assignments 1, 2, and 3 above? If the publications use abbreviated codes and symbols to save space, can they be read and understood easily? Are the identities and qualifications of reviewers described?

5 Choose some topic that interests you. Consult information sources selected for the above assignments and select two *titles* of each of the following media types: 16mm films, 35mm filmstrips, video tapes or discs, audio tapes or disc recordings, and supplementary books. Enter data on separate 4-by-6-inch cards for each item selected. Then locate a qualitative review of at least one of each of the media types. Note on the card for each item whether the review was favorable or unfavorable, and in what respects. If possible, preview, read, or audition one or more of the items, and compare your opinions with those of reviewers.

Helpful Hints

☐ Get acquainted with the card and book catalogs of your own college or university. Learn, too, whether some bibliographic information is available in microfiche or computerized data bank form, such as ERIC, DIALOG, MEDLARS, or the *Magazine Index*.

☐ When writing producers or publishers to obtain review copies of their catalogs and brochures, use official school stationery.

☐ Consult other information sources such as: *Educational Media Yearbook* (Brown and Brown); *Educators' Guide to Media Lists* (Sives); or *Guide to Reference Books for School Media Centers* (Wynar).

MOTION PICTURES: FILMS AND VIDEO

The Reason for It

At times, a motion picture on film, video tape, or video disk can be the most appropriate and effective teaching device available. But what the instructor does, and what the students do—before, during, and after a group showing—may make the difference between genuinely effective learning and simply passing time more or less pleasantly. Similarly, when a motion picture is assigned for independent study, the assignment or guide you supply your students is of critical importance. Utilizing motion pictures properly involves many professional decisions.

Before You Start

Brown, Lewis, and Harcleroad: *AV Instruction*, Chap. 11; Reference Section 6. This manual: Exercises 2, 9, 61, 62, 63; Correlated References, Sections 32, 41.

Purposes

1 To learn some of the unique contributions which can be made to learning by motion pictures in film or video form

2 To improve your skills in motion picture appraisal and criticism

3 To learn that teaching with motion pictures requires a careful consideration of educational purposes and a knowledge of their special contribution toward accomplishing these purposes

Required Equipment

1 A 16mm sound motion picture projector and screen; a video tape playback unit. NOTE: A video disc unit and one or more video disc subjects may be used, if available.

2 Catalogs and source lists describing available film and video subjects and their availability.

3 Selected film and video subjects.

Assignments

1 Examine various motion picture catalogs (especially those of firms that supply your own media center). Prepare a list of exact titles, sources, rental or purchase costs, running times, and producers of up to ten films or video productions that could be used advantageously in teaching subjects in which you are interested.

2 Select a 16mm film or video tape which seems appropriate for group teaching. View and appraise it, using the form provided.

3 Then prepare a written plan for teaching with the same film. Follow suggestions for group teaching provided in Brown, Lewis, and Harcleroad: *AV Instruction*, Chap. 11.

4 Finally, prepare for the students a written study guide to accompany the same film. Include specific questions for them to answer, and assignments to complete.

McGraw-Hill Films.

WORKSHEET 3

Name _____

Course _____ Date _____

YOU MAY COPY THIS FORM TO COMPLETE THE EXERCISE

Exact Title _____

Producer _____ Production Date _____

Medium   ☐ 16mm sound film   ☐ Video tape   ☐ Video disc

Running time _____ minutes   ☐ Black-and-white   ☐ Color

Accompanying material available: ☐ Teacher's guide   ☐ Lab exercises   ☐ Programmed worksheets

I. Appraisal

1   At what level of learning could this material be used effectively?

   ☐ Adult   ☐ Postsecondary   ☐ Secondary   ☐ Elementary

2   What is your overall evaluation of it?

   ☐ Superior   ☐ Excellent   ☐ Good   ☐ Little Value   ☐ Unusable

3   How do you rate the following? (Circle; 1 = excellent, 5 = poor)

Use of the visual capability	1	2	3	4	5
Use of motion, when needed	1	2	3	4	5
Use of voiceover	1	2	3	4	5
Use of synchronized sound	1	2	3	4	5
Technical quality of pictures	1	2	3	4	5
Technical quality of sound track	1	2	3	4	5

4   For what educational purposes would the item be best used?
   ☐ For learning one or more skills (psychomotor)
   ☐ For building or changing attitudes
   ☐ For developing appreciations
   ☐ For learning information (facts, principles, concepts)
   ☐ For a documentary, "slice-of-life" type of experience
   ☐ For entertainment

II. Planning for Use

1   What course objectives does this item fit? In what areas is it most useful?

2   What are the important content units (sequences, segments) of the item—in the order of their occurrence?

3   List words or phrases that need to be defined or described before using the item. What other preparation is desirable?

4   Describe at least three follow-up activities that might be good to ask students to do after using the item.

(Use additional sheets, if necessary.)

EXERCISE 4

BROADCAST AND CABLE TELEVISION

Broward County Schools, Florida.

The Reason for It

The ultimate medium of mass communication for both sight and sound, almost unlimited by space and time, is television. Television as an invention has no innate morality. As an instrument for learning, it depends upon the wisdom and integrity of its producers and its viewers. If the viewer is a student, a teacher may recommend programs. Subjects correlated with classroom lessons are many and varied; thus, the teacher must understand how to capitalize on television programs as instructional assets.

Before You Start

Brown, Lewis, and Harcleroad: *AV Instruction*, Chaps. 11, 12. This manual: Exercises 49, 64, 65; Correlated References, Section 41.

Purposes

1 To become familiar with currently available television programs and the televiewing habits of students

2 To analyze the values of a particular television program for in-school use or out-of-school assigned viewing

3 To plan the possible uses of an appropriate television program

Required Materials and Equipment

1 Weekly program listings of various television outlets: commercial, cable, community, public service, educational, or other

2 A television receiver

3 Study guides and program notes, if available

Assignments

1 Examine television program listings for a period of one week. List those programs which, from titles or program notes, seem to be of value to the grade or class you plan to teach. Select two programs specially designed for children or teens, one televised by an educational channel, the other presented by a commercial broadcast or cable station. Appraise each program, using copies of Form A, or design a form of your own.

2 Poll your class, another teacher's class, or a group of students of the age group you plan to teach. Ask each student to tell you or to write down the names of their three favorite programs. Tabulate the responses. Watch the resulting top favorite for two or three appearances on the air. Fill in Form B.

3 Compare your choices with those of students. What kinds of motivation might you supply to affect student choices? Comment on Worksheet Form C.

4 Locate organizations that provide evaluative reviews of television programming and that give special attention to programs of educational value. Names of organizations are included in such reference volumes as *Educational Media Yearbook*, Brown and Brown, Libraries Unlimited, Inc., Littleton, CO. (See section on Educational Media-Related Organizations: A Directory.) Obtain samples of their guides or bulletins to determine the information they provide that may help in finding and using out-of-school viewing opportunities for students.

5 Determine from your local program sources the existing policies about permitting schools to videotape broadcasts for subsequent use in class. How can you find out which programs may not be recorded under any condition, which may be recorded for restricted use, such as for a period of one week for a classroom or independent-study viewing only?

6 Make a list of the steps you would take to prepare your students for out-of-school viewing of some specific educationally valuable broadcast program. Where would you get background information for yourself and for them? How would you determine whether a broadcast should be recommended for viewing? How would you consolidate the benefits of the student viewing after the broadcast has been seen?

7 It is expected that over seventy percent of American homes will have cable television by the mid-1980s. This is an excellent source of local programs in many areas. Also, C-SPAN, short for Cable Satellite Public Affairs Network, broadcasts daily sessions of the House of Representatives. The Close-Up Foundation and C-SPAN carry question and answer sessions between high school students and political leaders. Both programs are valuable for social studies teachers. Study guides are available through C-SPAN, Tower Villas, 3800 N. Fairfax Dr., Arlington, VA 22203.

Name _____

Course _____ **Date** _____

YOU MAY COPY THIS FORM TO COMPLETE THE EXERCISE

Form A: Television Program Appraisal, Your Choices

Title of program viewed _____ Source _____

Date _____ Hour _____ Program length _____minutes

Criterion	Excellent	Good	Fair	Poor	Unsatisfactory
Content (authentic?)					
Content (useful information?)					
Content (otherwise not available?)					
Vocabulary (level suitable for your class?)					
Organization and continuity (appropriate?)					
Educational value (for your grade or class)					

What *key concepts* were emphasized in this program?

Comments:

Form B: Television Program Appraisal, Student Choices

1 Number of students polled _____

2 What is the *age range* of the group you have polled? _____ Average no. television sets in homes? _____

How many hours a week, on the average, does each student watch television programs? _____

What is the range of this viewing time for the group? _____

3 What is the *basic plot or format* of the top-rated program of the group you polled?

4 What *special techniques* used on the program seem to be especially appropriate and useful on television?

5. What *educational purposes*, if any, can be served by this program—from the point of view of the group you polled?

Form C: Comparison, Your Choices With Student Choices

1 Compare choices. Comment on type of program, educational value, time of broadcast.

2 Describe motivational materials and follow-up materials you could use to encourage student viewing of programs of your choice.

AUDIO RECORDINGS

San Diego County Schools, California.

The Reason for It

Much information is acquired either primarily or totally via the audio channel. Students of all ages readily identify with audio recordings. Radio, disc and tape recordings, and audio cassettes have made recorded materials immediately accessible to almost everyone. It is easy to incorporate these materials which are available from both educational and commercial sources into an instructional program. They carry subject matter for practically every field, including music, literature, historical and current events, sciences, languages, psychology, philosophy, politics, and government. With these materials, students can have inexpensive access to information that can be listened to almost anytime, anywhere.

Before You Start

Brown, Lewis, and Harcleroad: *AV Instruction*, Chap. 10. This manual: Exercise 52; Correlated References, Sections 3, 4.

Purposes

1 To enable you to locate—through the use of selection aids—audio recordings for educational applications

2 To prepare you to identify the distinguishing features and advantages of the several formats of audio recordings

3 To enable you to describe and design various learning situations in which audio recordings can be used

Required Equipment and Materials

1 Audio recording selection aids (see Correlated References)

2 Record player

3 Open reel audio tape recorder

4 Cassette audio tape recorder

5 Audio card tape recorder

6 Speech compressor record

Assignments

1 Select and evaluate at least two audio recording selection aids or catalogs in terms of scope and treatment, and the information provided. Use the form on the page following, "Audio Selection Aid Evaluation Form." You may copy this for your reports.

2 Using available audio recording selection aids and catalogs, develop lists of recordings applicable to one specific subject; to the degree possible from the information, separate your selections into recordings that can be used with large or small groups of students, and others for individual, independent listening.

3 From the list above, select a recording and prepare a lesson plan for using it: how to introduce it to a class, the activities that can be performed by the students during or after they hear the program, and possible further follow-up activities.

4 Select at least two instructional recordings, each in a different format and, using the "Audio Recording Evaluation Form" on the following page, evaluate each program. Explain how such a form (or the procedure used) can be of benefit to a teacher or group of teachers.

5 Use an audio card tape recorder. Select an audio card exercise from those supplied, and prepare a short description of typical instructional situations where audio cards would provide valuable learning experience. To determine the subjects treated in this audio format, explore the catalogs or selection aids that include audio cards.

6 Play a compressed-speech demonstration tape. Investigate the advantages and disadvantages of compressed speech as given in the literature. Are there occasions when speech reduced in speed to slower than normal might be advantageous? For what purposes would you consider using compressed speech?

Selection Aids

Index to Educational Audio Tapes, NICEM (National Information Center on Educational Media)

Index to Educational Records, NICEM (National Information Center on Educational Media)

Schwann Record and Tape Guide, Schwann Publications.

See also Correlated References 3 and 4, in this manual.

WORKSHEET 5A

Name _____

Course _____ Date _____

YOU MAY COPY THIS FORM TO COMPLETE THE EXERCISE

Title _____ Publisher _____

Frequency of Publication _____ Cost _____

1 Scope of selection aid. Relevance to education, comprehensiveness, adequate coverage of what is available:

2 Arrangement. Ease of use, concise arrangement, cross-indexes, indication of new entries, use of coding devices:

3 Treatment. Completeness of bibliographic data, quality of annotations, suggested uses, evaluations:

4 Format. Easy to use, handle, and read; conveniently laid out:

WORKSHEET 5B

Name _____

Course _____ Date _____

YOU MAY COPY THIS FORM TO COMPLETE THE EXERCISE

Title _____ Format ☐ Open reel tape   Speed _____
Producer _____   ☐ Cassette tape   (ips)

Artist _____   ☐ Disc   ☐ Stereo   ☐ Mono Speed _____
                                                                                    (rpm)
Available at _____ Playing time _____ minutes

1 Type of recording (underline appropriate words):   realistic   factual   fictional   dramatic   sentimental
emotional   documentary   historical   inspirational   informative   instructional   entertaining
philosophical   scientific   biographical

2 List instructional purposes or learning objectives that the recording is intended to (or could) achieve:

3 List topics covered:

4 Evaluate technical quality: (Excellent-E; Fair-F; Poor-P)
☐ Noise-free recording (no hum, scratches, irrelevant sounds)
☐ Speech sounds clear, pleasant, and well delivered
☐ Mechanical condition of the tape or disc satisfactory
☐ Fidelity (quality of recording; distortion)

5 Instructional goals to which the recording might contribute:
☐ Learning factual information
☐ Learning principles or concepts
☐ Learning operating procedures
☐ Learning motor skills
☐ Developing attitudes, appreciations, values or opinions

6 Overall evaluation:
☐ Excellent
☐ Fair-Acceptable
☐ Unsatisfactory

14

EXERCISE 6

FILMSTRIPS

Edmonton Schools, Alberta.

The Reason for It

Name almost any subject, and you should be able to find an up-to-date filmstrip about it. Though some filmstrips may have printed captions on each picture (frame), most have an accompanying tape or disc recording with a narration, indigenous sounds, sound effects, and music. Some sound filmstrips have audible cues indicating when to change frames, while others have inaudible pulse cues that change frames automatically in appropriate projectors. Filmstrips are flexible and convenient to use; they may be projected for groups or viewed by individuals. Presented in a fixed sequence, visual materials provide structure for subjects studied. But, as with motion pictures and all other media, a teacher needs to learn how to select the right filmstrips, for use at the right time, in the right situation, and with appropriate utilization for maximum student benefit.

Before You Start

Brown, Lewis, and Harcleroad, *AV Instruction*, Chaps. 4, 9; Reference Section 1. This manual: Exercise 43; Correlated References, Section 24.

Purposes

1 To provide experiences that will enable you to acquaint yourself with filmstrips, their physical characteristics, how to project them, and their usefulness in education

2 To acquaint you with sources, content, and values of several filmstrip source lists and producers' catalogs

3 To assist you in improving your skill in selecting, evaluating, and planning to use filmstrips related to your teaching subject and grade level

Required Equipment and Materials

1 A sound filmstrip projector.

2 Source books which list filmstrip titles from various producers and supply descriptive data about the content and treatment of the filmstrips. One example of such a source list is the NICEM *Index to 35mm Filmstrips.*

3 Catalogs from educational filmstrip producers.

4 A number of periodicals which contain qualitative reviews of filmstrips recently produced.

5 Several 35mm sound filmstrips with available study guides.

6 If available, a motion picture on how to utilize filmstrips in instruction.

Assignments

1 If available, review the sound motion picture, *Children Learn from Filmstrips* (McGraw-Hill Films), or other audiovisual materials describing values and uses of filmstrips.

2 Review the available source lists, commercial catalogs, university or school media-center catalogs, noting productions related to your subject field and grade level interests. Prepare a list of at least seven sound filmstrips that appear to be useful to you. Include on your list all of the data you would need in requesting your school media center staff to obtain them for review or purchase.

3 Classify the filmstrips under three categories: (a) highly useful, (b) moderately useful, and (c) of no value. Cite reasons for your ratings.

4 Project and review at least three sound filmstrips and read through study guides that may accompany them. From these filmstrips, select one title for use in completing the worksheet that follows.

Orange Unified School District, California.

WORKSHEET 6

Name _____

Course _____ Date _____

YOU MAY COPY THIS FORM TO COMPLETE THE EXERCISE

Title _____ Producer _____

Sound ___ Silent ___ No. of frames ___ Color ___ b and w ___ Production year ___ Study guide? Yes ___ No ___

I Appraisal

1 At what learning level could this material be used effectively? Check one or more if appropriate:

☐ Adult ☐ Postsecondary ☐ Secondary ☐ Elementary

2 What is your overall evaluation?

☐ Superior ☐ Excellent ☐ Good ☐ Of little value ☐ Not usable

3 For what type of group would the filmstrip be most appropriate?

☐ Large (auditorium) ☐ Large (classroom) ☐ Small (2–10) ☐ Individual study

4 How do you rate the filmstrip on the following five-point scale? (1 = superior; 5 = poor)

Are the pictures appropriate and of good quality?	1	2	3	4	5
Does it raise questions?	1	2	3	4	5
Is it accurate and up to date?	1	2	3	4	5
Does the sequence do a good job of structuring the topic?	1	2	3	4	5
If a sound filmstrip, does the sound track supplement the pictures?	1	2	3	4	5

5 Describe briefly the content and any spoken words or visual concepts which should be explained in advance of use.

II Planning for Use

1 What course objectives does this filmstrip fit?

2 Describe briefly how you would use this filmstrip in a particular class. How would you introduce it?

3 What key questions should the students consider and answer after viewing the filmstrip?

4 In what activities might students appropriately engage while viewing the filmstrip?

5 What additional student activities might be generated by seeing the filmstrip?

6 Describe briefly how you might evaluate the outcome of showing this filmstrip to your class.

EXERCISE 7

SELECTING AND USING FLAT PICTURES

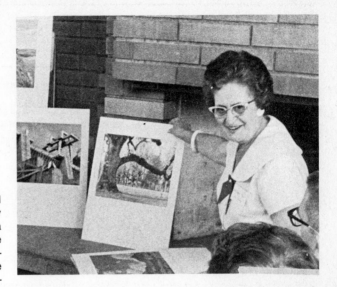

The Reason for It

To read flat pictures successfully, students need special skills. Reading even simple pictures with these skills, they can generate hundreds of written or spoken words and a wide variety of learning activities. Teachers need to be able to choose useful flat pictures for most areas of the curriculum, for many instructional objectives can adequately be achieved by most students only if they have access to pictorial resources. For every teacher, the ability to choose useful flat pictures is basic, and is a skill that has transfer value when choosing other media.

Before You Start

Brown, Lewis, and Harcleroad: *AV Instruction,* Chap. 9; Reference Section 1. This manual: Exercises 31, 42; Correlated References, Sections 39, 40.

Purposes

1 To provide experience that will help you develop skill in selecting flat pictures and illustrations

2 To provide experience in the use of the opaque projector

Required Materials

1 Some flat pictures obtained from free or inexpensive sources, such as magazines or sponsored pamphlet materials. Choose pictures which have potential value for teaching in your special area.

2 Inexpensive mounting board, rubber cement (or dry-mounting tissue and iron), and binding tape.

3 Opaque projector, screen, and room with light control.

San Diego County Schools, California.

Assignments

1 Number the free pictures you have selected (up to five), analyze them, and answer the questions about them which are asked on Worksheet Forms A and B.

2 Study the criteria on Worksheet Form C. Star on the worksheet those criteria which a busy teacher should consider to be most important in choosing flat pictures.

3 Analyze your pictures according to the four criteria on Worksheet Form C; rate each one on the worksheet, and determine whether each one should be used in teaching.

4 If possible, arrange your flat pictures in a useful sequence. Mount them on uniform-sized pieces of mounting board, using rubber cement or dry-mounting tissue. Use binding tape to join the mounts together so the pictures may be projected sequentially by an opaque projector or used as a chalkboard or corner-table display. See further details in Exercise 11. Project them and see if the picture details and sequence are appropriate.

5 Prepare a brief plan for teaching with the flat pictures you have selected and mounted. Consider the following matters:

• The *proper sequence* in which to display them

• What *questions to ask* or *comments to make* about each

• What special *words* or *terms* might need to be explained about each

• Ways in which you could use each one in the process of *evaluating student learning*

WORKSHEET 7

FORMS FOR ANALYSIS OF FLAT PICTURES

Name _____

Course _____ **Date** _____

YOU MAY COPY THIS FORM TO COMPLETE THE EXERCISE

Form A Prepare titles for all five (or more) of the free pictures you have selected, entering them in the proper spaces below. Use the numbers 1 to 5 in completing the remainder of this assignment.

No.	Picture Title	Source of Picture	Rating from Form C
1			
2			
3			
4			
5			

Form B What *abstractions* can be made real with each of the above pictures?

												Abstractions
No.	Sounds	Temperature	Motion Speed	Size	Distance Depth	Mass, Weight	Odor, Taste	Feel, Touch	Color	Emotion	Time
1											
2											
3											
4											
5											

Form C *Evaluate* each of the free pictures (by number), using the "excellent" to "unsatisfactory" ratings below, and indicate on Form A whether each one should be used.

Factor	Excellent	Good	Fair	Poor	Unsatisfactory
Interesting and attractive. Catches the eye and holds it.					
Large and simple, so it may be seen clearly.					
Important Information for topic being studied.					
Accurate information. Truthful; up to date; shows comparative size.					

EXERCISE 8

STORING AND DISPLAYING FLAT PICTURES

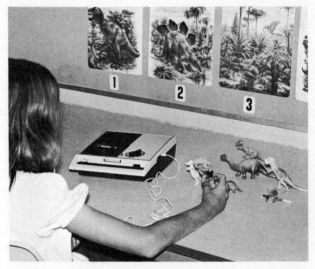

The Reason for It

Flat pictures have numerous uses in teaching: to focus the attention of an entire group on some recognizable scene, to provide a basis for testing or asking questions, to encourage students to talk about what they see, to provide illustrative materials for bulletin-board displays. To have flat pictures readily available for such teaching purposes requires that they be cataloged systematically and filed in ways that permit them to be retrieved quickly when needed. Having an efficiently arranged file, you may be encouraged to display your pictures in innovative ways.

Before You Start

Brown, Lewis, and Harcleroad: *AV Instruction,* Chaps. 9, 16; Reference Section 4. This manual: Exercises 7, 10, 23, 24, 42. Correlated References, Sections 39, 40.

Purposes

1 To develop a system for obtaining, cataloging, and filing flat pictures

2 To produce several devices used to display flat pictures for instructional purposes

Required Equipment and Materials

1 A supply of magazines from which pictures may be clipped

2 A supply of 9- by 12-inch manila file folders or suitable substitutes

3 A cardboard box of appropriate size to hold folders

4 Miscellaneous supplies, such as coat hangers, stiff cardboard suitable for making display easels, cloth or plastic tape

5 An opaque projector

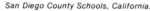
San Diego County Schools, California.

Assignments

1 Choose some topic of interest that might be included in or comprise an entire unit of work—China, wheat farming, transportation, or pollution, for example.

2 Search through various expendable sources of flat pictures from which you are free to clip. (See especially sources listed in various publications identified in Chapter 16, Free and Inexpensive Materials in *AV Instruction: Technology, Media, and Methods.)*

3 As you locate and clip promising pictures, organize and label separate file folders for them. Use ordinary 9- by 12-inch standard manila file folders, or make your own. Arrange the files and the pictures within each file in such a way as to be of maximal value to you in your teaching. For some uses, you will want to mount clipped pictures. To file mounted pictures, use separators cut from file folders as shown. Revise or add to file classifications as necessary, taking care to keep the filing system simple and workable—convenience encourages use!

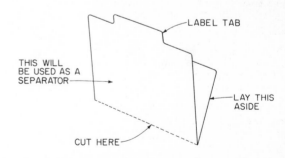

4 Select pictures from ten different file classifications and write a brief guide for using each picture. Include: (a) some questions you might ask students about the picture, (b) some ways you, the teacher, might display or use the pictures, and (c) some ways the students might use the picture.

5 Gain experience in using an opaque projector to display flat pictures. Also study and follow suggestions described in Exercise 6 for enlarging pictures or drawings by use of the opaque projector. Be sure to observe precautions against harming your pictures through prolonged exposure to projector heat.

6 Select a group of pictures that form a suitable sequence for an "accordion" display, as pictured below. Write a study guide to accompany the display, giving suggestions for using it, questions to be asked about pictures, and related materials.

Helpful Suggestions

Illustrated below are several ways in which to display the pictures you have selected, cataloged, and filed for your collection. Can you invent more ways that are equally useful?

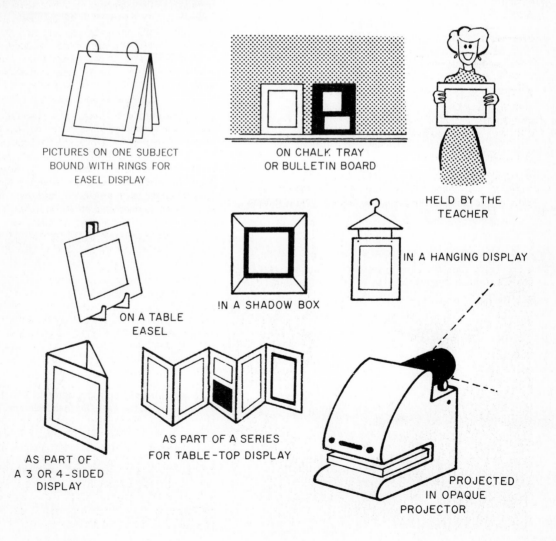

PICTURES ON ONE SUBJECT
BOUND WITH RINGS FOR
EASEL DISPLAY

ON CHALK TRAY
OR BULLETIN BOARD

HELD BY THE
TEACHER

ON A TABLE
EASEL

IN A SHADOW BOX

IN A HANGING DISPLAY

AS PART OF
A 3 OR 4-SIDED
DISPLAY

AS PART OF A SERIES
FOR TABLE-TOP DISPLAY

PROJECTED
IN OPAQUE
PROJECTOR

REFERENCE MATERIALS

The Reason for It

Most libraries now maintain comprehensive media centers and a staff to assist researchers, teachers, and students in locating and obtaining materials needed for research and study activities. All of these individuals need access to articles and reports to support their analysis of subjects. Selection aids and processes used in assignments for this exercise will extend your acquaintance with reference materials and research reports, and improve your skills in using them.

Before You Start

Brown, Lewis, and Harcleroad: *AV Instruction,* Chap. 17; Reference Section 6. This manual: Correlated References, Sections 1, 2.

Purposes

1 To acquaint you with the contents of selected basic reference materials and how to use them

2 To provide experiences by which you will learn how to use sources that offer information about exemplary educational programs and about educational research and development efforts

Required Materials

There are a number of widely used reference sources mentioned in the assignments in this exercise and others are cited in the Correlated References section of this manual.

FLAT PICTURES   C 31 *FRS*
FILMSTRIP
SHOWS HOW TO USE FLAT BOARDS EFFECTIVELY. FROM THE HOW TO FOR CHURCH SCHOOL TEACHERS SERIES.
LC NO. FIA65-2421
PROD-CONCOR   *DIST*-CONCOR   1964

FLAT PICTURES   C 52 *FRS*
SOUND FILMSTRIP - RECORD   T
PRESENTS SIX DIFFERENT WAYS TO USE FLAT PICTURES. FIVE TECHNIQUES ON TEACHING WITH FLAT PICTURES. THE USE OF QUESTIONS WITH PICTURES AND REASONS WHY AT TIMES FLAT PICTURES ARE MORE EFFECTIVE THAN MOTION PICTURES. FROM THE TEACHING WITH INSTRUCTIONAL MATERIALS SERIES.
LC NO. 79-737278
PROD-OLM   *DIST*-OLM

FLEA   C 99 *FRS*
SOUND FILMSTRIP - AUDIO TAPE   J-H
SEE SERIES TITLE FOR DESCRIPTIVE STATEMENT. FROM THE CHECKERED FLAG SERIES B SERIES.
LC NO. 78-737693
PROD-FEP   *DIST*-FEP   1969

FLECTCHER MARTIN   C 53 *FRS*
SOUND FILMSTRIP - RECORD   J-H A
FLETCHER MARTIN EXHIBITS OIL PAINTINGS AND DEMONSTRATES HIS TECHNIQUES AS HE WORKS IN HIS STUDIO. FROM THE FAMOUS ARTISTS AT WORK SERIES.
LC NO. 79-735415
PROD-SCHLAT   *DIST*-SCHLAT   1969

FLETCHER AND ZENOBIA   C 60 *FRS*
SOUND FILMSTRIP - RECORD   K-P
USES THE STORY OF THE SAME TITLE BY VICTORIA CHESS AND EDWARD GOREY TO FOCUS ON THE LINK BETWEEN THE IMAGINATION AND THE REAL WORLD. PRESENTS THE STORY, COMBINED WITH PARTICIPATION GAMES, RHYME AND OTHER DEVICES, TO LEAD THE CHILD FROM OBSERVATION OF THE WORLD AROUND HIM TO AN EXTENSION OF THAT WORLD THROUGH FANTASY. FROM THE OBSERVING AND IMAGINING SERIES.
LC NO. 70-736955
PROD-PARENT   *DIST*-LCOA   *PRODN*-TSE 1970

FLIGHT - OVERCOMING DISTANCE   C
SOUND FILMSTRIP - RECORD   I-C
SEE SERIES TITLE FOR DESCRIPTIVE STATEMENT. FROM THE WORLD WAR I TO TOMORROW SERIES.
PROD-INCC   *DIST*-LCOA

FLIGHT INTO SPACE   C 70 *FRS*
FILMSTRIP WITH CAPTIONS   I
EXAMINES THE THIRD LAW OF MOTION VIA THE EXPERIMENTAL APPROACH, THE FIRST AND SECOND LAWS OF MOTION, FRICTION AND GRAVITY AND THEIR AFFECTS ON THE PROPULSION OF A BODY THROUGH THE ATMOSPHERE, THE PROBLEMS INVOLVED IN A ROCKETS ATTAINING ORBITAL VELOCITY AND THE VAST SETUP BEHIND SPACE FLIGHT. FROM THE MACMILLAN ELEMENTARY SCIENCE FILMSTRIPS SERIES.
LC NO. 75-735859
PROD-MACM   *DIST*-MACM   *PRODN*-INSTMP 1970

FLORENCE NIGHTINGALE   B 37 *FRS*
NICEM.

Assignments

1 Review and become familiar with the contents and organization of several of the following publications which provide general information about different types of reference sources:

Cheney, F. N.: *Fundamental Reference Sources*
Katz, W.: *Introduction to Reference Works, Vol. 1: Basic Information Sources*
Sheehy, E. P.: *Guide to Reference Books*
Wynar, B.: *Best Reference Books*

If some of the titles are not available, look for similar publications in the list of background material for this exercise in the Correlated References.

2 In the library or media center, locate recent monthly copies and cumulated volumes of *Education Index* and *Library Literature.* These two publications list articles in periodicals dealing with all phases of education and with resource materials to keep researchers and teachers in step with new developments. Also consult magazines related to other subject fields which are similarly indexed in services such as the *Reader's Guide to Periodical Literature* and *Applied Science and Technology Index,* and automated data bases such as DIALOG, ORBIT, and MEDLARS. If necessary, ask the librarian or media specialist for help in locating or using them.

3 If the library you use subscribes to the services of ERIC (Educational Resources Information Center), investigate the nature of its offerings. Learn how to use its microfiche collection, if available; review the contents, organization, and usefulness of ERIC-related publications such as: *Resources in Education* (RIE), *Current Index to Journals in Education* (CIJE), and the *Thesaurus of ERIC Descriptors.* Ask the librarian for other ERIC documents that may be on hand. Assess ways in which the various ERIC services may be of help to you in your teaching.

4 Utilize a recent issue of the *Education Index* or *Library Literature* or the *Automated Magazine Index* to locate one article dealing with "television" and another with "censorship." Use information given in the index (author of article, title, magazine in which published, volume number, page numbers, and date of publication) to locate the two periodicals you selected. At the top of a 4- by 6-inch card, place complete bibliographic information for each article. Follow this with a brief annotation of the article content.

5 Secure from your library a recent monthly or cumulative issue of *Resources in Education.* Check the subject index section under the heading: Reading. Select one publication listed under this heading. Provide the librarian with the ERIC identifying number. The publication you receive will be in either printed or microfiche form. After reading it, summarize important points in outline form.

6 Address a request to the ERIC Clearinghouse in Information Resources, 130 Huntington Hall, Syracuse University, Syracuse, N. Y., 13210, asking to receive an information packet about the ERIC system. Investigate whether ERIC services are available to you in the libraries or media centers in the region where you work.

7 If you are teaching in an elementary or secondary school choose two encyclopedia sets that appear to be best suited to needs of students at the grade level and in the subject field of your interest. Proceed as follows: For information on appraisals of encyclopedia sets, study the guidelines for reviewing encyclopedias appearing in Cheney's *Fundamental Reference Sources;* consult Sheehy's *Guide to Reference Books* and Katz' *Introduction to Reference Works.* With the above information as a background, locate in your library the two sets to which you then narrow your choice; examine

each carefully, using Cheney's guidelines as a point of reference; finally, state reasons supporting your final choice.

8 In preparation for this assignment, consult Cheney, Sheehy, Wynar, and Katz, and, using procedures recommended in Assignment 7, choose two atlases appropriate to the requirements of your students.

Helpful Hints

• Remember that periodical literature often provides the most up-to-date information about educational developments. To facilitate your search for articles in current magazines, encourage your library or media center to subscribe to the *Education Index, Library Literature, Magazine Index,* or *Current Index to Journals in Education.* These publications have monthly supplements that are cumulated on a yearly basis.

• Investigate the possibilities of using computer-based data banks to locate additional resources pertaining to subjects of professional interest, of which the ERIC, DIALOG, ORBIT, and MEDLARS referred to in this exercise are examples.

ERIC Central.

EXERCISE 10

FREE, SPONSORED, AND INEXPENSIVE MATERIALS

The Reason for It

The curriculum is often enriched by use of quantities of free, sponsored, and inexpensive information and promotion materials such as booklets, maps, charts, reports, storybooks, films, and filmstrips. Also, many free materials that students bring from home or obtain from local businesses serve valuable purposes—for example, junked clocks, old motors, ice cream containers, and display stands. Some sponsored and inexpensive materials are designed especially for student use; others are not. Those designed for adults may sometimes be adapted for younger students. But the fact that these materials may be free, or inexpensive, is not the primary concern; the chief criterion is: Do they make worthwhile contributions to learning?

Before You Start

Brown, Lewis, and Harcleroad: *AV Instruction,* Chap. 16. This manual: Correlated References, Section 25.

Purposes

1   To learn, firsthand, about the characteristics of a broad sample of free, sponsored, and inexpensive materials

2   To become acquainted with a number of widely used source books and compendiums that provide information about the availability and content of these materials

3   To learn to develop and apply criteria in selecting free, sponsored, and inexpensive materials

Required Materials

1   Reference copies of several of the source lists and compendiums listed in the Correlated References, Section 24, for this exercise. See others listed in Brown, Lewis, and Harcleroad, *AV Instruction,* Chap. 16.

2   Several samples of free, sponsored, or inexpensive materials that relate to your teaching interests, obtained from representative sources.

Assignments

1   Select one of the free materials source lists mentioned in Correlated References, Section 24, or in Brown, Lewis, and Harcleroad, *AV Instruction,* Chap. 16. Read annotations and select at least two organizations to which to write for materials. Follow directions in Helpful Hints below.

2   When the requested materials arrive, appraise one of them, using the form "Free Material Appraisal" on the following page.

3   Bring your item to class to become part of a cooperative display of free, sponsored materials. Attach to your item full information as to its source and possible educational use. Be prepared to comment about the qualitative values and detractions of sponsored free materials for use in teaching. NOTE: You may prefer to obtain your item in person through visiting local offices of organizations mentioned in your source list.

4   Using only scrap materials that are free and commonly available from homes of students, construct some device or display that can be used to teach some concept or to develop some skill in a subject field of your choice. Try it out with one or more students of about the age of those with whom you would intend to use it. From your observations, draw conclusions as to its educational worth.

Helpful Hints

Here are several suggestions to aid you in obtaining sponsored free materials:

• If you are now teaching, write on official school or institutional stationery. If not, make clear in your letter that the request grows out of an assignment for your class.

• Be clear as to your expectations of receiving the item without charge, including handling and postage.

• Give all details in your first letter. Mention the grade level at which you wish to use the material, the topic being studied, and intended uses of what is requested.

• If you have addresses for both the main office and a branch office of the organization supplying the material, write only to one.

• Do not expect to receive large quantities of materials. Justify requests for more than one copy.

• Write directly to sources of materials, *not* to the publisher of the source list itself.

• Be sure to retain copies of all letters sent.

WORKSHEET 10

Name _____

Course _____ **Date** _____

YOU MAY COPY THIS FORM TO COMPLETE THE EXERCISE

Directions: Carefully examine the free item you have obtained as a result of carrying out Assignment 1. Comment on (appraise) it with respect to each of the following characteristics:

1 *Freedom from undesirable or excess advertising.* If an item does contain advertising, is it in good taste? Is it sufficiently limited in amount? If not, can it be limited through editing?

2 *Importance.* Does the item present useful information that relates directly to one or more important instructional goals? If so, describe the goal(s) and evaluate extent of application of the item to achieving it.

3 *Presentation effectiveness.* Is the item designed and written to communicate effectively—i.e., without ambiguity, and with the use of familiar terms and an approach that is suited to the students with whom you might use it?

4 *Usefulness.* To what specific educational purposes could you put the item in instruction? For each such use, what specific contributions would you expect the item to make?

24

EXERCISE 11

TEXTBOOKS

The Reason for It

A frequent prediction is that the electronic revolution, to which so many of the media described in this manual are related, will reduce drastically the need for textbooks and similar types of printed materials. It is our belief that this will not be the case. Our prediction is that increasing use will be made of systematically designed textbooks and other printed instructional materials, often as parts of multi-media kits or packages. As a teacher, you must, therefore, be able to examine and appraise textbooks and text-type materials and to use appropriate sources of data to aid you in recommending those that meet criteria imposed by community mores or by local school district or state educational guidelines.

Before You Start

Brown, Lewis, and Harcleroad: *AV Instruction,* Chap. 17; Reference Section 6. This manual: Exercises 12, 16, 19, 51; Correlated References, Section 10.

Purposes

1 To provide information regarding criteria typically applied to the selection and evaluation of textbooks

2 To suggest sources of information pertaining to those criteria

3 To offer experience in applying those criteria to the evaluation of three specific textbooks

Required Equipment and Materials

1 Three textbooks intended for use in the same subject and at the same grade level

2 Several reference sources that contain evaluative and technical review data about the texts selected

Assignments

1 Skim through each of the three textbooks you have selected. Note the general appearance, content, and organization. At various points throughout each, insert slips of paper upon which you have made notes of your observations and reactions.

2 Check the *Education Index,* the *Book Review Digest,* the Curriculum Advisory Service's *Curriculum Review,* Bowker's *Textbooks in Print,* and publications of the Educational Products Information Exchange (EPIE) and the Association of American Publishers, such as: *How To Get The Most Out of Your Textbook* and *Textbook Questions and Answers,* or similar sources noted in the Correlated References section to locate qualitative or technical reviews of your three textbooks.

3 With results of Assignments 1 and 2 as references, perform a detailed analysis of the three textbooks and evaluate them, using the form provided.

WORKSHEET 11

TEXTBOOK EVALUATION FORM

Name _____

Course _____ Date _____

YOU MAY COPY THIS FORM TO COMPLETE THE EXERCISE

Study thoroughly the three texts you have just skimmed; now evaluate them in depth by answering each criterion question listed in the form below with a Yes or No. Before answering the questions below, carefully follow the instructions for each which are stated in parentheses.

Names of texts

BASIC CONSIDERATIONS IN EVALUATING TEXTBOOKS

CONTENT
Is the author (or authors) regarded as competent in the field? (Check the writer's background in a reference such as *Who's Who in America.*)

Does the text interpret the curricular objectives of the subject as prescribed by the course of study? (Select a subject area in the text and compare information given in the book with the objectives set up in a course of study for a similar subject.)

Does the text contain biased views on controversial issues? (Read sections dealing with such areas as: use of natural resources, development of nuclear power, role of women in society, air pollution, sexuality.)

Does the subject matter appear to promote sound moral values and contain writing of literary merit? (Read several random paragraphs and, as you read, note such features as sentence structure, use and choice of colorful words and objectionable slang expressions.)

TREATMENT OF CONTENT

Is the style suitable for the age level of the students for which it is intended? (Check such items as length of sentences, length of paragraphs, logical and suitable comparisons.)

Is the vocabulary suitable for the grade it is intended for? (Check the words used on two or three pages with a basic word list for the grade for which book is intended.)

Does the material adapt itself to individual differences? (Examples: Will the material appeal to girls and to boys? Are helps included for slow readers? Are there extra challenges for superior students?)

Is there evidence that women, children, and minorities are not stereotyped and are portrayed in roles of leadership and centrality when such portrayal is reasonable? (Check photographs and sketches as well as written content.)

Does the content delineate life in contemporary urban environments as well as in rural and suburban settings?

ARRANGEMENT OF CONTENT

Are the index and table of contents complete and easy to use? (Note the subjects mentioned on one or more pages in the text. Check these subjects with the entries in the index. If the index is to be really useful, most of the subjects should be included in the alphabetic arrangement.)

Are difficult and unusual words included in a glossary?

Are illustrations, maps, sketches, tables, graphs used to supplement the printed matter? (Select any ten pages in the text and tabulate the number of pages that have no illustrations, only one, more than one.)

Do the visuals add interest to the textual matter? (Check for such factors as reality in color, artistic page arrangement, size ample for good perception, minimum of irrelevant details.)

Are the suggested related activities practical, and do they add information not given in the text? (Analyze the activities to determine amount of time, special facilities, out-of-school resources, etc., needed to carry out each activity.)

Do well-organized summaries and reviews appear at the end of chapters and units?

Do the bibliographies include the most up-to-date materials, both printed and audiovisual?

Does the publisher provide media such as films, filmstrips, recordings, and transparencies to supplement the book?

VERIFICATION

Does the publisher provide evidence that, prior to publication, the textbook was tested with students of certain characteristics and that it was revised and improved, as necessary?

Does the publisher provide further evidence that, since publication, students of specified characteristics who have used the textbook have learned satisfactorily?

MECHANICAL STANDARDS

Is the type clear and plain?

Is there enough leading, or spacing between lines, to make the text easy to read? (Four inches of vertical space should include no more lines than the number indicated for each age level: under 7 years, 10 lines; 7 to 9, 20; 9 to 12, 22; and above 12, 24 lines.)

Are the lines of proper length for easy reading (not more than 4 inches nor less than 3 inches in length)?

Is the paper of good weight and durability?

Is the binding reinforced so that the book is held firmly in the cover?

Are the pages planned for easy readability? (Check the following factors: pages inked evenly, ample margins, nonsmear ink, etc.)

Is the text available in both hardbound and paperback?

EXERCISE 12

PROGRAMMED MATERIALS

The Reason for It

One approach to learning through self-instruction involves the use of programmed learning materials. The concept of programmed learning is based upon a process that enables an individual student to progress successfully through a structured sequence of steps toward objectives specified in behavioral terms. The student is self-paced, kept actively responding, guided toward making correct responses, and given immediate confirmation of progress. When programmed learning is used selectively in modularized instruction and similar programs, students can achieve objectives with economy of time and reach a high level of proficiency.

Before You Start

Brown, Lewis, and Harcleroad: *AV Instruction,* Chap. 17. This manual: Correlated References, Section 10.

Purposes

1 To locate programmed materials and to identify them as different from other forms of instructional materials

2 To recognize the components of programmed instruction, and to be able to distinguish between linear and branching formats

3 To develop effective ways to integrate appropriate programmed materials into instructional practices

4 To evaluate programs and their role in meeting educational needs of students

Required Materials

Samples of linear and branching programs in printed and audiovisual formats for different age groups and different subject areas.

Assignments

1 Consult Carl Hendershot's *Programmed Learning Bibliography,* and list half a dozen programs suited to a specified group of students according to grade level and subject matter.

2 Learn about the characteristics of programmed materials by:

a Reading in such references as Cram, David D.: *Explaining "Teaching Machines" and Programming,* Fearon Publishers, Belmont, Calif., 1961; Markle, Susan: *Good Frames and Bad* (2d ed.), John Wiley, New York, 1969; and Pipe, Peter: *Practical Programming,* Holt, Rinehart and Winston, New York, 1966.

b Viewing such films as *Programming is a Process: An Introduction to Instructional Technology* (U. of Illinois); *Programmed Instruction: Developmental Process* (National Audio Visual Center); or the filmstrip *Teaching Machines* (Basic Skill Films).

3 Read Mager on *Preparing Instructional Objectives;* then write a sequence of three behavioral objectives with criterion test items which could be used as a basis for a short program in your area of interest.

4 Using the form "Appraising and Using Programmed Materials" on the next page as a guide, work through a program of interest to you. Complete the form, and, in addition, develop an outline of ways the program could be integrated into an instructional activities sequence.

program patterns

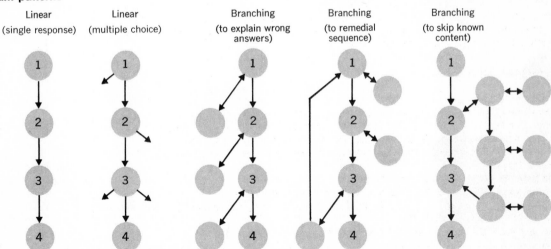

Linear (single response) — Linear (multiple choice) — Branching (to explain wrong answers) — Branching (to remedial sequence) — Branching (to skip known content)

Types of Programs

The broad types of programmed materials which have been developed can be classed under two general headings: linear or branching.

Most linear programming is structured to have students write or construct responses so that they will progress correctly through small steps or tasks toward achieving specified objectives. Each response students make is reinforced immediately according to the principles of operant conditioning as described by B. F. Skinner. All students complete the items or frames in the same linear sequence at their own pace.

Branching programming is designed to enable students to progress through a sequence of frames according to their experience, competence, and interests. They are presented with an instructional statement which could consist of one or two paragraphs; then they are given a related problem to solve. After completing the task, they select from alternative answers provided in the program, and if their answers are faulty, they are guided into further instruction or branched into frames that will provide the additional needed background information and appropriate examples. This technique is not as much concerned with producing "error-free" learning (Skinner) as with enabling students, through the use of diagnostic questions, to proceed according to their needs.

Each type of program can be produced in either written or mediated forms. There are several variations and combinations of the two approaches to help meet diverse needs.

WORKSHEET 12

APPRAISING AND USING PROGRAMMED MATERIALS

Name _____

Course _____ Date _____

YOU MAY COPY THIS FORM TO COMPLETE THE EXERCISE

Title of program _____ Publisher _____ Author(s) _____

Subject_____ Grade Level_____ Date_____ Validated: Yes/No

1 *Objectives*

   a Are they behaviorally stated? Yes/No

   b Is a criterion test included? Yes/No

   c Describe how the program relates to the educational needs of the students:

2 *Program*

	Good	Fair	Poor
a Step size	_____	_____	_____ (too large/complex)
b Level of difficulty	_____	_____	_____ (too simple/hard)
c Frame sequence	_____	_____	_____ (confusing)
d Cuing, prompting techniques	_____	_____	_____ (insufficient)
e Response relevance	_____	_____	_____ (meaningless)

3 *Utilization*

Describe how the program will be used in regard to:

   a Teacher preparation

   b Student preparation

   c Support activities and materials required

4 *Evaluation*

State the evaluation procedures you will use to determine the effectiveness of the program relative to learning and attitudes.

EXERCISE 13

COMPUTER PROGRAMS: SELECTION AND USE

The Reason for It

To do their work, computers require programs—*software,* a series of instructions or codes. Furthermore, computer services are only as good as the programs they process. Selection of those programs is an essential teacher responsibility.

Before You Start

Brown, Lewis, and Harcleroad: *AV Instruction,* Chap. 15. This manual: Exercises 13, 67. Correlated References, Section 13.

Purposes

1 To become familiar with three major types of microcomputer programs for use in instruction.

2 To analyze values of computers as aids to learning.

3 To obtain experience in planning possible uses of computers and related software in the classroom.

4 To learn how to use an evaluation procedure in the selection of computer programs.

Required Materials and Equipment

1 Microcomputers on which to run and evaluate programs. Note that programs run only on designated computers. It is advantageous if you can use the same make and models of microcomputers that you will have to use in your work assignments. Seek the machines you need: try schools, computer stores, college and university laboratories, area or regional education centers.

2 Programs of assorted types suitable for subjects and levels of interest to you. Again, search the sources listed in 1, above.

3 Catalogs, listings from publishers or manufacturers, lists in publications on educational computing to find leads on programs appropriate for your needs.

Assignments

1 Select a topic and grade level in which you have teaching experience, or plan to teach in the future. Using available computerware catalogs, computer store listings, advertisements, or your own institution's collections, prepare a list of programs that appear to relate to what you are interested in.

2 Run up to three of the programs you have selected. If you do not know how to do this, seek assistance from your instructor, from others in the class, your school's computer coordinator, or individuals in local computer stores.

3 Using the worksheet on the following page, apply criteria to the programs you have selected.

4 Compare your evaluations with those of others in the class. Can you arrive at any generalizations regarding them? Any recommendations?

5 Talk with teachers who have used microcomputers in their classrooms. What reactions and suggestions do they have regarding them?

Control Data Corporation.

Helpful Hints

• *Teacher-prepared programs.* These require some programming knowledge and some command of programming language (usually BASIC—"*B*eginners' *A*ll-purpose *S*ymbolic *I*nformation *C*ode").

• *Commercially-prepared programs.* Those completed by other educators. They may be run by teachers or students who have no programming knowledge.

• *Author language programs.* These provide teachers with a format, or structure, of a lesson and require them to fill in their own questions, answers, or problems. These do not require programming knowledge or skill, but users must understand rules and codes employed.

Minicomputer programs are usually available as cassette tape recordings or discs. They are nearly always accompanied by a set of instructions or a manual to assist users.

WORKSHEET 13

COMPUTER PROGRAM APPRAISAL FORM

Name ______________________________

Course ____________________ **Date** ____________________

YOU MAY COPY THIS FORM TO COMPLETE THE EXERCISE

Content area selected ______________________________

Grade level selected ______________________________

Program 1

Name ______________________________ Author/Producer ______________

Program 2

Name ______________________________ Author/Producer ______________

Program 3

Name ______________________________ Author/Producer ______________

Rating Scale: (1) Excellent or Good; (2) Fair; (3) Poor

Criteria	Ratings/Comments		
	Program 1	**Program 2**	**Program 3**
Appropriateness of *content* for your students			
Degree to which program makes effective use of the *interactive mode*			
Appropriateness of the program *vocabulary*			
Clarity and understandability of program and manual *instructions*			
Effectiveness of program *graphics* (if used)			
Accuracy of information presented in program			
Overall rating and comments:			

EXERCISE 14

SELECTING GAMES AND SIMULATIONS

Close-Up; C-SPAN.

The Reason for It

Educational games and simulations can make a valuable contribution to classroom learning. Well-designed games, involving appropriate participation activities, can ensure dynamic learning experiences for students of any educational level. Younger students may profit most readily from playing games with simple materials and structured rules; even drills in basic learnings such as mathematics and spelling can be practiced in a game format. In secondary and postsecondary education, sophisticated game/simulations can be used in studies of economics, political science, sociology, human relations, and advanced problems in the sciences, mathematics, and complex decision-making exercises in international relations and commercial enterprises. However, successful use of games and simulations depends on careful selection and thorough testing of their educational values in actual practice. Since many professional decisions are involved, a study of games and simulations is warranted.

Dr. Marjory Golick and Jeffrey Norton Publishers, Inc.

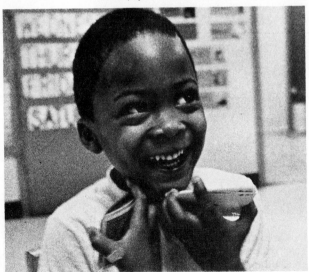

Before You Start

Brown, Lewis, and Harcleroad: *AV Instruction,* Chap. 14. This manual: Exercise 39; Correlated References, Section 26.

Purposes

1 To learn some of the unique contributions that games and simulations can make to learning

2 To develop some skills for appraising the educational values of games and simulations

3 To learn some techniques of teaching with games and simulations relating such use to accepted educational purposes

Required Materials

1 Catalogs of games and simulations

2 Selected games of various types: card games, board games, manipulative games, and games for reading, math, and science

Assignments

1 Examine catalogs of games and simulations in your library or media center. List several games/simulations that appear to be of value in your teaching field.

2 Check out several games/simulations, analyze them, and play any that appeal to you.

3 Using the worksheet on the following page, analyze two of the games that you believe to be most useful for the instructional program in your interest field. Complete the forms with a brief lesson plan for using each game/simulation.

United States Department of the Army.

ARCADE ENGAGEMENT SIMULATOR

WORKSHEET 14

SELECTING GAMES
AND SIMULATIONS

Name _____

Course _____ Date _____

YOU MAY COPY THIS FORM TO COMPLETE THE EXERCISE

Appraisal Form—Game Simulation

Title_____

Producer _____ Cost _____

Subject matter area _____

Suggested grade level _____

1 Are rules and directions clear, direct, short? Comment.

2 Does teacher's manual discuss teacher's role? Comment.

3 Comment on appearance, construction, packaging.

4 Supplementary materials. What is needed? Are parts of the game consumable? Costs?

5 Playability: How much involvement is there? Is there much idle time? Are rules too complex? Too easily violated? Interest level? Comment.

6 To what extent does material meet curricular need? Comment.

7 Develop below a short plan for using this material for teaching. Identify learning objective(s), grade level, and utilization plan.

8 Is the game self-checking? Or is supervision required?

Appraisal Form—Game Simulation

Title_____

Producer _____ Cost _____

Subject matter area _____

Suggested grade level _____

1 Are rules and directions clear, direct, short? Comment.

2 Does teacher's manual discuss teacher's role? Comment.

3 Comment on appearance, construction, packaging.

4 Supplementary materials. What is needed? Are parts of the game consumable? Costs?

5 Playability: How much involvement is there? Is there much idle time? Are rules too complex? Too easily violated? Interest level? Comment.

6 To what extent does material meet curricular need? Comment.

7 Develop below a short plan for using this material for teaching. Identify learning objective(s), grade level, and utilization plan.

8 Is the game self-checking? Or is supervision required?

EXERCISE 15

MAPS AND GLOBES

The Reason for It

Seeing the world through the symbolic medium of maps and globes requires highly developed skills on the part of both teachers and students. Sound judgment in selecting and teaching with maps and globes requires an understanding of map symbolism, scales, and distortions. In short, understanding how maps communicate is a subject that warrants study by teachers and students.

Before You Start

Brown, Lewis, and Harcleroad: *AV Instruction,* Chap. 6. This manual: Exercise 27; Correlated References, Section 30.

Purposes

1 To learn some of the values of maps and globes in instructional programs

2 To analyze and compare values of a specific globe and a commercially produced map or set of map transparencies

3 To learn the many inexpensive sources of maps, methods for using them, and what teachers can expect students to learn from working with maps and globes

4 To use maps and globes in the evaluation of learning, at the various levels of instruction

5 To provide practice in planning the use of these materials

Required Materials

1 A commercially made globe of the world

2 A set of commercially made flat classroom maps designed for use at the level you teach or will teach, and appropriate for the subjects in which the maps will be used

3 A selection of free or inexpensive maps that are used locally such as road maps of states, regions, or cities. These may be purchased or obtained from some service stations, government agencies, chambers of commerce, or stationery and map stores

4 Map transparencies, as available

Assignments

1 Review the maps available to you, and decide which are most likely to be of value in your program.

2 Using the appraisal portion of the worksheet—next page—evaluate the items selected, according to the qualities listed.

3 In Section 4 of the worksheet, make a plan for the use of one of the maps you have evaluated and judged satisfactory.

4 Using the map or globe you have selected, devise a plan for student self-testing of map-reading ability. Try to make the test reveal what students know about map reading, and include means for them to identify what more they need to learn to master reading at a satisfactory level.

WORKSHEET 15

Name _____

Course _____ **Date** _____

YOU MAY COPY THIS FORM TO COMPLETE THE EXERCISE

Title _____ Globe_____ Map_____

Source _____ Size _____ Type of mount _____

1 *Description of content presented in the item.* (Underline appropriate words: physical, political, relief, historical, social, economic, other; then describe content in more detail below.)

2 *Type of map projection used?*

3 Using the 5-point scale below, *appraise the above item* according to the criteria presented.

	Excellent	Good	Fair	Poor	Unsatisfactory
Grade-level-content suitability					
Legibility for entire class					
Legibility for individual study					
Appropriate color symbolism					
Appropriate and accurate detail					
Ease of handling					

4 *Brief plan for use of map, map transparency, or globe:*
specific objectives to be achieved:

Classroom activity of teacher and students:

Specific method of measuring achievement of students. Include additional learning materials if needed for independent study.

34

EXERCISE 16

MULTI-MEDIA KITS

The Reason for It

Multi-media kits provide a convenient way to introduce variety into learning and to accommodate instruction to individual student needs and interests. Kits are available from many producers of instructional resources. But, to meet local and group requirements, teachers often produce them, sometimes with the help of students. Either commercial or locally made kits usually include several different media, for example, audio cassettes, slide units, flat pictures, and various types of printed materials, study guides or syllabi, printed participation materials—sometimes in the form of reproducibles, pamphlets, and brochures. Some kits contain games, sample materials, construction materials, or devices for practical experiences. Among skills most important to teachers is the ability to select or to assemble and produce multi-media instructional kits.

Before You Start

Brown, Lewis, and Harcleroad: *AV Instruction,* Chaps. 2, 4, 17. This manual: Exercises 12, 51; Correlated References, Sections 27, 31.

Purposes

1 To become familiar with the contents of a few typical multi-media kits and criteria for selecting them

2 To learn to plan and develop original media kits, and to combine existing materials to make kits that will provide students with experiences by which they may achieve planned objectives

Required Materials and Equipment

1 A suitable commercial multi-media kit that includes at least two types of media and a study guide

2 Miscellaneous commercial media items appropriate to objectives to be taught (Assignment 2)

3 Equipment as needed to use items in the kits selected for study and designed for production

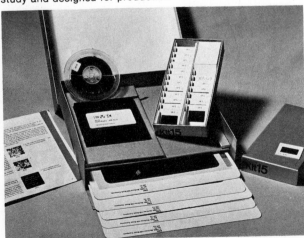

Assignments

1 Choose a commercially prepared multi-media kit for a subject field and teaching level of interest to you. Appraise it, applying criteria suggested in the worksheet, following, for Assignment 1; add and use other criteria you think should be included.

2 Plan and produce a multi-media kit. Use existing and available commercially produced materials, or create some of your own. Your kit should include at least three types of materials, a syllabus or guide that describes the objectives the kit is designed to serve, recommended ways to use the kit materials, and means of evaluating outcomes from study with the kit materials. Choose or develop materials such as audio tapes, slides, graphics, games, 8mm films, transparencies, and printed or duplicated resources. Complete the form for Assignment 2 (following page).

Helpful Hints

Well-planned multi-media kits include a simple, convenient procedure for making an inventory of the contents, thus reducing the danger of losing items. Three inventory systems are in common use, and others can be devised:

• *Matching method.* A drawing of the kit layout is affixed to the lid, and location of items is indicated by print, code numbers, or the shapes of the section drawings.

• *Color code method.* Each item in the kit is of a different color or is color-coded and is fitted in a section colored to match. Colors may be applied with paint or with commercial adhesive-backed color stickers or labels, or patches of adhesive-backed shelf paper may be used. For each kit and each kit unit, different color combinations and different symbols will distinguish individual items and their proper locations.

• *Number inventory method.* Each kit item carries an identifying code number that tells which kit the item belongs to and the number of the individual item. A reference list in the kit box lid makes the inventory check easy.

WORKSHEET 16

**APPRAISING AND PLANNING
MULTI-MEDIA KITS**

Name _____

Course _____ Date _____

YOU MAY COPY THIS FORM TO COMPLETE THE EXERCISE

Assignment 1. Appraising a Commercially Produced Multi-Media Kit

Title of kit_____

Producer or publisher_____

Production date _____

1 Prepare a descriptive list of each item in the kit.

2 List the objectives for which the producer intends the kit to be used; state your opinions as to the usefulness of the kit to students who are trying to achieve those objectives. Give reasons to support your opinions.

3 Evaluate the teacher's guide that accompanies the kit in terms of: (a) practicality of utilization suggestions; (b) evidence of prerelease testing of the kit and results reported; (c) educational philosophy of the kit designers, as you can infer from the teacher's guide.

4 Evaluate the study guide that accompanies the kit in terms of: (a) readability; (b) practicality of assignments; (c) correlation with kit contents—i.e., does it contain all materials that are needed to complete the assignments? If not, are they readily and economically available through local sources? Are appropriate tests provided to evaluate student achievement?

5 Evaluate the system used to inventory the kit contents. Suggest improvements.

Assignment 2. Planning Your Own Multi-Media Kit

Curriculum topic_____

Grade level _____ Kit title _____

Objectives to be achieved with the kit:

Items contained in the kit and source of each: Equipment required?

1

2

3

4

5

Plan for utilizing the kit (describe):

EXERCISE 17

USING COMMUNITY RESOURCES FOR LEARNING

The Reason for It

Because there is a gap between most activities within schools and those in the world outside the school, many students mature with inadequate understanding of the complexities and responsibilities of adult life. To help bridge this gap between the sheltered, often isolated school program and life in the adult community, many schools make a strong effort to use community resources for learning. This exercise, and the one following (Field Trips), are intended to encourage broadening the areas of school activity to include many resources from the community.

Before You Start

Brown, Lewis, and Harcleroad: *AV Instruction,* Chap. 3; This manual: Exercise 18; Correlated References, Section 12.

Purposes

1 To prompt you to explore areas in your community suitable for student experience and study

2 To explain how a community resources survey can be developed

3 To identify activities in the community that students might correlate with their educational interests and their future needs for understanding the world of work, recreation, and government

Lawrence Hall of Science, Berkeley.

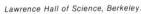

Assignments

1 *Community resources files.* Determine whether your college or school, or regional educational service agency maintains community resources files for teacher use. If you can locate such a file, make a list of questions about it and prepare a report of your findings. In your report include answers to such questions as: Is the file up to date? For what school levels does it appear to be designed? How is it organized: By curriculum topics? By kinds of community enterprises? What data are included in statements about each community resource? Are people—individuals—listed as community resources? Are there instructions on how to use each listed resource? Are there recommendations for instructional objectives that might be realized by using each listed resource? Are limitations and regulations stated? Does the record include names, addresses, telephone numbers, data about when school participation is welcome? Does the information include statements about opportunities for students to work free or for pay? Are there indications of materials, publications, or other media that can be made available to students? Are there suggestions for preparatory work that students should undertake before approaching the community resource agency? What other information is included? NOTE: By all means, teachers should be aware of local school board policies for using community resources.

2 *Schools using the community as a learning center.* Determine whether there is a list of schools that regularly provide community resource study activities for their students. If you find such a list, or learn of schools with such programs, make arrangements to visit one of them and interview the person designated to tell about the program. Obtain any announcements of the program, or printed reports about it. And, if you can, interview one or more students participating in the program to learn student reactions to it: Benefits? Problems? Advantages and disadvantages? What is actually done in the community by the students? For what recognition or rewards? What records are kept of student performance and contributions?

3 *Topics as leads to resources.* Make several lists of topics about which to build a community resource file for yourself. Think of these topics as a means of focusing on the use of community resources for learning. Here are some examples of topics: businesses, industries, transportation (air and ground); health, safety, government, laws and justice; garbage and trash disposal and recycling, water supply; youth services, services for the elderly; communication: public, private, local, intercity, interstate, and international. As you make the list, note areas of instruction where students could tie school work to an agency outside, either by active study or by service work. Make a plan for matching individual student interests to agencies you have identified, and then explore possibilities of having students become involved in studies and activities outside school. Remember to work with your school counselors, teachers, supervisors of work-study programs, other teachers, and your administrators. In your plan, identify people who are interested in helping youth to become valuable, employable, and constructive adults. And, at your first opportunity, start — even in a small way — to initiate activities in which students use community resources for worthwhile learning experiences.

4 *Locating local resources: agencies.* Use the yellow pages of your local telephone directory to locate main and branch libraries, main museums and neighborhood museums, art galleries, zoos, aquariums, planetariums, parks, historical spots and buildings, and recreation centers. When you have a list, select one or two places that seem most directly related to the interests of your students; investigate these agencies and learn what they have to offer students at different educational levels. Are they sites for only educational field trips, or individual visits, or do they have programs of special classes, lectures, or film showings? Do they provide employment opportunities for volunteers or for part-time workers such as docents?

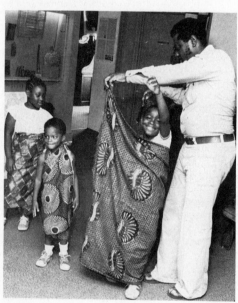

Lawrence Hall of Science, Berkeley.

5 *Locating local resources: people.* Make a list of people, and sources of contacts with people, who can bring to the school their experiences and talents, and share their enthusiasms with students. Some, of course, will prefer to have students visit them. Will they consent to being recorded? If so, have students plan to make recordings of such people — either on audio tape and with slides or Polaroid pictures, or with a portable video tape system. Develop a program of using people as resources in various aspects of instruction; design the opportunities for students to ask questions of these people and to observe them at their work. Share your list with other students in your class, or with other teachers, and ask them to share their knowledge of resource persons with you. Place a copy of your list of resource visitors in the instructional resources center for others to use and encourage them to add to it.

6 *Alternative schools.* Check in your community to determine whether alternative schools are in operation. Especially check to find those that provide community experiences for their students. Some may provide for a student to spend at least half time in community organizations or companies, learning and rendering services as an employee. Some may program almost all of a student's time to be spent outside a school environment, with occasional seminars or checkups with teacher-supervisors. When you locate an alternative school that uses the community extensively as a learning center, prepare a report on its program, and identify practices that might be adopted easily by other schools. Identify also problem areas in the program; such problems might be in the area of evaluation of student accomplishment and progress, or perhaps in your opinion certain assigned student activities may not be of educational value or within range of student abilities or interests. Attempt to determine what corrective measures might be taken to improve the effectiveness of the program, but use and disseminate such information judiciously.

Helpful Hints

• Remember, in using community resources for learning, that many demands are made upon you and your students to be diplomatic, cooperative, and thoroughly dedicated to the benefits of the activities undertaken. A gram of courtesy will outweigh the many inevitable blunders and natural mistakes made by students as they learn.

• Again, on courtesy: The gracious verbal thanks by students and by yourself for the assistance and cooperation of organizations and individuals — all given on a pleasant, person-to-person basis and confirmed in even a brief note of appreciation — will win support for your program and will open opportunities for other students.

• Any use of community resources for learning should be backed by instructional planning no less detailed or carefully designed than any other important instructional activity. Establish objectives, work out the arrangements and the necessary supporting resources, and summarize and evaluate. Reports of the evaluations should be shared with cooperative organizations and individuals who also should have participated in the planning of the student activity.

San Diego County Schools, California.

EXERCISE 18

FIELD TRIPS

The Reason for It

One way for students to explore the local community and to use it as a learning center is for them to participate in planned field trips. Because a field trip is a complex enterprise, the decision to undertake one requires careful prior assessment of the importance and benefits of the probable results. This exercise will give you an opportunity to analyze criteria for selecting a field trip site, procedures to ensure a smooth and productive trip, and means for evaluating the outcomes of the experience.

Before You Start

Brown, Lewis, and Harcleroad: *AV Instruction,* Chap. 3. This manual: Exercise 17; Correlated References, Section 12.

Purposes

1   To locate field trip sites that have high potential for contributions to the instructional program

2   To assess the values of a trip and the means by which optimum results from the project can be achieved

3   To devise a complete plan for a field trip, including necessary advance arrangements, procedures for activities on the site, and follow-up activities

Required Materials and Equipment

1   A guide to field trips or community resources prepared by a school district (see Exercise 17), preferably one nearby

2   Transportation to and from a selected field trip site

3   One or more means of recording events and details connected with the trip, such as materials for taking notes; portable audio tape recorder; 35mm, Polaroid, or Colorburst camera; or portable video tape recorder

Assignments

If you are a preservice teacher:

1   Identify a curriculum unit or subject that you may be expected to teach. From goals specified, select those that might include objectives that could be advantageously achieved by taking a field trip.

2   When you have selected these goals, study the field trip or community resource guides in the curriculum library of

San Diego County Schools, California.

San Diego Zoo, California.

**PLEASE DO NOT
ANNOY, TORMENT,
PESTER, PLAGUE,
MOLEST, WORRY,
BADGER, HARRY,
HARASS, HECKLE,
PERSECUTE, IRK,
BULLYRAG, VEX,
DISQUIET, GRATE,
BESET, BOTHER,
TEASE, NETTLE,
TANTALIZE, OR
RUFFLE THE ANIMALS**
SAN DIEGO ZOO
SAN DIEGO WILD ANIMAL PARK

your institution or of the local school district(s) to locate one trip that you believe to be of probable value. Using the worksheet on the page following, appraise the trip from the information in the published material.

3   If it is possible to do so, arrange to visit the field trip site you have selected to analyze, and using a copy of the form, again appraise the field trip site from your own observations. Compare the results of your two evaluations. Would you consider the trip justified in terms of your goals? Was the printed analysis of the trip accurate, clear, and useful as a guide? If not, in what ways was it inaccurate?

If you are an in-service teacher:

1   Select a field trip site that could be useful for your students. Obtain any information you can about experiences of others who have visited the same location. Check the files of your school or district media center for study guides about the location or trip reports from prior years. Were their goals similar to yours? What were the results of their trips? Did they make any suggestions or report problems to be avoided? Compare these findings with criteria for evaluation of potential field trips. If everything looks good, take the next step.

2   Make a personal visit to the field trip site. Analyze your findings on the worksheet. If you observe places, events, or activities in progress that may be helpful in introducing the field trip to your students, take notes, or pictures, or make audio recordings; but do not present so much to them that their motivation is dulled; use materials that motivate interest and help them to know what they will have opportunity to learn, and how it may be of importance to them.

3   (Project) Using the experiences listed above as background, prepare a plan for conducting a field trip to the location you have assessed. Include in your plans such information as:

• Recommended advance preparations by teacher and students.

• Materials and equipment to use on the trip.

• Recommended follow-up activities and courtesies.

• Materials to be prepared, including: permission sheet, for parents to sign and return, with information about the trip and the objectives for it; a map for the students; statements of en route observations to be made; vocabulary or special terms; plan for the trip record, including equipment to be used; and special assignments for individual students. In some instances a shooting script for still or motion picture photography may be prepared. Finally, prepare a test or other means of evaluating learning from the trip.

WORKSHEET 18

Name _____

Course _____ **Date** _____

YOU MAY COPY THIS FORM TO COMPLETE THE EXERCISE

Use the following form to record essential information about a potential field trip site or locale.

Destination_____ Address_____ Date of visit _____

Check: Data derived from: ☐ Study guide   ☐ Other teachers   ☐ On-site visit

1 Subject area(s) for which trip is useful:

2 Objectives that can be achieved from this trip:

3 What supplementary study materials does the agency make available for use?

a. In the classroom

b. On the trip

4 What special contributions will this trip make for students?

a. Things they can see

b. Things they can do, especially by using cameras and tape recorders or video tape recorders

c. Important concepts that should be better understood as a result of taking the trip

5 What special classroom study is necessary before making the trip?

6 What special arrangements are necessary to make the trip successful?

a. Safety precautions

b. Time and travel arrangements

c. Finances—admission or transportation cost

d. Behavior standards

e. Trip permit slips required to visit this location?

f. Equipment needed, such as maps, cameras, tape recorders, or portable video tape recorders

7 What other learning materials are available to assist in the preparatory or follow-up work?

8 Are the special contributions of this field trip worth the time, effort, and preparation? Yes_____ No_____ Why?

EXERCISE 19

SELF-INSTRUCTION FACILITIES AND EQUIPMENT

Dallas Community College.

The Reason for It

With the increasing amount of student self-instruction in many programs, design of appropriate study spaces is of great importance. Basic requirements for independent study include a comfortable and adequately furnished area as well as instructional media resources and equipment. The types of facilities that are available are almost infinite in variety. Instructional planners must be aware of alternatives and know how to select the most feasible and appropriate.

Before You Start

Brown, Lewis, and Harcleroad: *AV Instruction*, Chaps. 2, 3. This manual: Exercises 1, 19; Correlated References, Sections 22, 23, 28, 31.

Purposes

1 To become acquainted with a facility for self-instruction

2 To develop a basis for choosing types of equipment, materials, space, and facilities most suitable for self-instruction

Assignments

The general assignment is to select, visit, observe, and evaluate a self-instructional program. From a wide variety of locations, choose one that seems to have probable value for you to investigate. Here are a few suggestions: schools or colleges, of any level, and programs of instruction in any subject field; business, industry, or government programs; public libraries, continuation centers, hospitals, store buildings, or retirement centers. The facilities may be located in a comprehensive media center, in a classroom or laboratory, in special rooms, or in corridors, outdoors, in parks, or on factory lands.

When you have selected your locale for observation, make these preparations and complete the assignment:

1 List briefly the reasons for your selection of the locale and what you want to learn there—the goals for your visit.

2 Prepare a checklist of questions for your investigations, including items that are likely to apply where you intend to teach.

3 Decide how you may want to present your report—in writing or in a visual presentation or a combination of both. Select your equipment: for example, cassette recorder for interviews to obtain quotations for reporting in print or on an edited audio tape; Polaroid camera for quick, full color paper snaps for inclusion in a written report; 35mm camera for slides for a slide-sound report.

4 Prepare to answer the questions on your checksheet for either a written or a visual presentation.

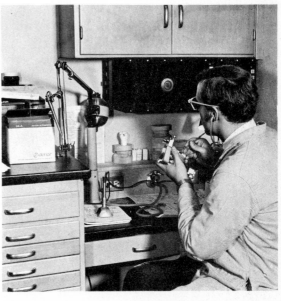

Horace Hartsell.

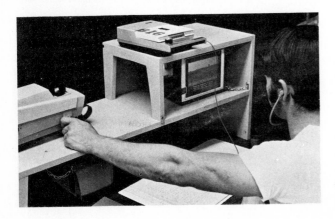

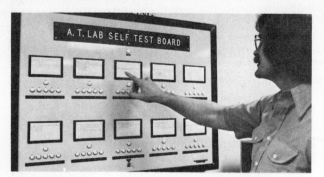

Portable Media for Self-Instruction

There seems to be no need to await elaborate and sophisticated technical systems to start work with media to serve independent student study. As schools attempt to individualize instruction, existing audiovisual equipment may be the most conveniently available and economical means to test and evaluate teaching and study procedures and the materials needed for the program. Simple, portable equipment is provided in most schools; setting up a convenient environment in an appropriate location to facilitate the use of the equipment is a first step toward media utilization.

Self-Instruction in Study Carrels

The possibility of immediate student access to audio and to visual experiences in semipermanent or permanent facilities for self-instruction challenges the imagination. The systems shown here are representative of a bridge between the use of existing items of equipment and their associated media, and the systems—dial-access, remote-access, and computer-based systems—now under exploratory development. As a stage of development toward more sophisticated technology, carrels may have 120-volt power outlets for serving portable equipment units of many kinds, and a simple distribution system using microphone, coaxial cables, or audio lines leading from the carrels to a central master control may provide adequate program distribution.

Many people have supported far-out systems, with complete contact between students in carrels and a bank of program sources, with many diverse lessons available by turning a dial. While many practical systems are now available and some are in use, the dream of having every necessary resource for study of any subject at the fingertip of each student is not likely to be realized in the immediate future. Perhaps the delay is desirable. In the meantime, both teachers and students can—with immediately available audiovisual and printed resources—learn the advantages and the techniques of developing instruction for self-directed, self-activated study.

Lawrence Hall of Science, Berkeley.

San Diego County Schools, California.

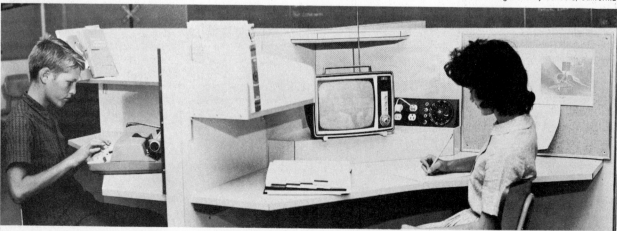

SECTION TWO

CREATING INSTRUCTIONAL MATERIALS

Creating instructional materials is a process that helps students to learn. Sometimes it is the very best activity through which learning objectives can be achieved. It is especially valuable in learning facts, procedures, or concepts that are difficult to master. The process of creating communication resources often helps to clarify problems and leads to enriched learning from the production effort itself.

Values in Creating Materials

The chief value of creating instructional resources is the *learning experience* the activity provides for students who plan, produce, and use them as means of communicating clearly with others. Such productions are not do-it-yourself means of reducing school budgets. Rather, they are experiences that help students become expert in communicating ideas visually and dramatically. Busy-work assignments should be avoided; projects should have value to the students who create them; originality and inventiveness should be encouraged.

Levels of Construction

The creation of instructional resources may be carried on at several successively more exacting and thoughtful levels. For the sake of convenience these levels may be classified as (1) imitative, (2) adaptive, and (3) creative. Each is a different kind of learning experience. Two students creating the same thing may actually be performing at two different levels. The three levels are not rigid categories; rather, they are stages in a continuum ranging from simple projects complete with how-to-do-it directions to complicated projects that require student initiative, research, experimentation, and evaluation.

☐ **Imitative Level** At the imitative level, emphasis is upon following directions. On their own, many students enjoy following directions to build models, to sew, or to engage in other construction projects. In school, a student may be given a set of directions and asked to produce a map or model. In addition to following directions, the student is expected to maintain performance and neatness standards, to be accurate and prompt, to use appropriate knowledge and skills in carrying out assignments, and to improve in his capacity for self-direction.

☐ **Adaptive Level** At the adaptive level, individual judgment and initiative assume added significance. Here there are no set directions to follow, and, although prototypes may exist, no one of them completely satisfies requirements. The student is thus asked to investigate, to choose and try various alternatives, and to use known principles, materials, tools, and techniques in seeking solutions to the creative problem. Students who wish to construct a scale model of an Elizabethan stage, for example, are soon involved in a study of many different aspects of English theater. They may locate pictures of stages of this period that will help. But to produce an accurately scaled model they must bring into play other skills, including those of library research, mensuration, scale and proportioning, art, and design. The finished product represents their adaptations of data and facts as they discovered and interpreted them through their investigations.

☐ **Creative Level** At the creative level, stress is upon original thinking of a problem-solving nature. Although clues for solutions to design or content problems may be found to exist in already created devices, instruments, or processes, students in this case are unfamiliar with them and are re-

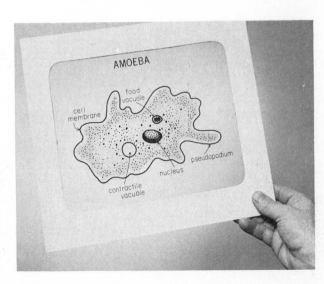

quired to create them in their effort to solve the instructional problem. In doing so they follow usual problem-solving procedures to (1) isolate the problem, (2) develop ideas offering promise for its solution, (3) select likely alternatives to test, (4) evaluate results of each attempt, and (5) draw conclusions about the best or most nearly valid solution. An additional concern, from the viewpoint of this section, is the production of a concrete device or material that can be used to communicate the findings and their bases to other interested persons. As an example of creative invention, a student wishes to find a means of demonstrating the electrical conductivity and resistance of various sizes of copper wire of uniform lengths and metallic content. After several tries he develops a large board on which he mounts several 50-watt light bulbs, each connected in parallel to a uniform length of covered copper wire of varying thickness. When the panel is connected to a source of power, resulting contrasts in brilliance of the different bulbs provide visual evidence of the relationship of electrical resistance to size of conductor. The student investigator thus solves the problem and, at the same time, communicates his findings to others.

Recipes for Construction

Several of the construction activities described later in this manual require the use of plastic mixtures. Some of these can be made in the classroom, using the recipes below:

☐ **Puppet mix.** Ingredients: ½ cup table salt; ½ cup cornstarch; ¼ cup water. Mix ingredients together thoroughly, then cook over low fire, stirring continually. Material quickly stiffens into a lump. When it is sufficiently cooked, knead it briefly. It is then ready to use. To color, add dissolved watercolors, melted crayons, or ink to original mix, or paint it when it is dry. The product can be wrapped in waxed paper and stored in refrigerator for future use.

☐ **Papier-mâché.** Can be made in several ways: (1) soak torn paper bits in thin paste and mix well; (2) boil paper bits; mix them until they form a smooth mass; squeeze out water, add paste, glue, and plaster of paris; (3) tear toilet paper into shreds, boil, and beat until smooth, squeeze out water, add paste; (4) dip inch-wide strips into paste and lay over torn or wadded paper center to produce desired form.

☐ **Plaster of paris.** Select an enameled container which will clean easily after use. Pour in the amount of water needed (about half the volume of the finished mixture). Without stirring, slowly sift plaster of paris into the water; continue as long as it sinks. When enough is added so that the powder stays on top without being absorbed in the water, stir and press out lumps by hand. Use mixture immediately, before it dries.

☐ **Flour and salt mixture.** Use 2 cups of flour and enough water to mix into creamy consistence. Add 1 cup of salt. Hardens when dry.

Section Two Exercises

After first examining issues related to copyright law (in Exercise 20), the processes of creating many different types of instructional resources are presented. Assignments for thirty-one exercises are made in units on lettering; bulletin boards and displays; chalkboards; graphs and time lines; clothboards and magnetic boards; flat pictures; spirit and electrostatic duplicating; handmade transparencies; picture lifts; diazo and heat-process transparencies; games and simulations; puppets; audio tapes; visual literacy, including filmstrips without cameras; still pictures; 2-by-2-inch slides; motion pictures (film and video); visualized testing; and self-paced learning modules. With each of them, certain fundamental skills will be learned. Many times, skills learned in one exercise will be used again, in refined form, in completing another.

COPYRIGHT LAW

The Reason for It

The Copyright Law of the United States (revised) took effect January 1, 1978. This new law gives certain exclusive rights to the copyright owner; it also provides certain other rights to educators and students who wish to use copyrighted works. Simply having equipment capable of copying materials does not convey any right to do so. Knowing the conditions under which you may or may not copy is your professional responsibility. Each decision concerning whether or not to copy is important, because, no matter how little, unauthorized copying is unfair to publishers and authors. Actually, it is *stealing!*

Purposes

1 To become familiar with the principal exclusive rights of copyright owners

2 To become familiar with and to be able to apply certain limitations on exclusive rights through "fair use"

Assignments

1 Read the following excerpts from Circular R-21, *Reproduction of Copyrighted Works by Educators and Librarians* (Washington, Library of Congress, 1978).

2 Take the self-check quiz on the following page; check your answers. NOTE: Please understand that the quiz and the answers are to be considered as a guide only; they have no legal status.

NOTE: The following is a reprint of the entire text of section 107 of title 17, *United States Code.*

§ 107. Limitations on exclusive rights: Fair use

Notwithstanding the provisions of section 106, the fair use of a copyrighted work, including such use by reproduction in copies or phonorecords or by any other means specified by that section, for purposes such as criticism, comment, news reporting, teaching (including multiple copies for classroom use), scholarship, or research, is not an infringement of copyright. In determining whether the use made of a work in any particular case is a fair use the factors to be considered shall include—

(1) the purpose and character of the use, including whether such use is of a commercial nature or is for nonprofit educational purposes;

(2) the nature of the copyrighted work;

(3) the amount and substantiality of the portion used in relation to the copyrighted work as a whole; and

(4) the effect of the use upon the potential market for or value of the copyrighted work.

NOTE: The following is an Agreement on Guidelines for Classroom Copying in Not-for-Profit Educational Institutions

WITH RESPECT TO BOOKS AND PERIODICALS

The purpose of the following guidelines is to state the minimum and not the maximum standards of educational fair use under Section 107 of H.R. 2223. The parties agree that the conditions determining the extent of permissible copying for educational purposes may change in the future; that certain types of copying permitted under these guidelines may not be permissible in the future; and conversely that in the future other types of copying not permitted under these guidelines may be permissible under revised guidelines.

Moreover, the following statement of guidelines is not intended to limit the types of copying permitted under the standards of fair use under judicial decision and which are stated in Section 107 of the Copyright Revision Bill. There may be instances in which copying which does not fall within the guidelines stated below may nonetheless be permitted under the criteria of fair use.

GUIDELINES

I. *Single Copying for Teachers*

A single copy may be made of any of the following by or for a teacher at his or her individual request for his or her scholarly research or use in teaching or preparation to teach a class:

A. A chapter from a book;

B. An article from a periodical or newspaper;

C. A short story, short essay or short poem, whether or not from a collective work;

D. A chart, graph, diagram, drawing, cartoon or picture from a book, periodical, or newspaper;

II. *Multiple Copies for Classroom Use*

Multiple copies (not to exceed in any event more than one copy per pupil in a course) may be made by or for the teacher giving the course for classroom use or discussion; *provided that:*

A. The copying meets the tests of brevity and spontaneity as defined below; *and,*

B. Meets the cumulative effect test as defined below; *and,*

C. Each copy includes a notice of copyright.

Definitions

Brevity

(i) Poetry: (a) A complete poem if less that 250 words and if printed on not more than two pages or, (b) from a longer poem, an excerpt of not more than 250 words.

(ii) Prose: (a) Either a complete article, story or essay of less than 2,500 words, or (b) an excerpt from any prose work of not more than 1,000 words or 10% of the work, whichever is less, but in any event a minimum of 500 words.

[Each of the numerical limits stated in "i" and "ii" above may be expanded to permit the completion of an unfinished line of a poem or of an unfinished prose paragraph.]

(iii) Illustration: One chart, graph, diagram, drawing, cartoon or picture per book or per periodical issue.

(iv) "Special" works: Certain works in poetry, prose or in "poetic prose" which often combine language with illustrations and which are intended sometimes for children and at other times for a more general audience fall short of 2,500 words in their entirety. Paragraph "ii" above notwithstanding such "special works" may not be reproduced in their entirety; however, an excerpt comprising not more than two of the published pages of such special work and containing not more than 10% of the words found in the text thereof, may be reproduced.

Spontaneity

(i) The copying is at the instance and inspiration of the individual teacher, and

(ii) The inspiration and decision to use the work and the moment of its use for maximum teaching effectiveness are so close in time that it would be unreasonable to expect a timely reply to a request for permission.

Cumulative Effect

(i) The copying of the material is for only one course in the school in which the copies are made.

(ii) Not more than one short poem, article, story, essay or two excerpts may be copied from the same author, nor more than three from the same collective work or periodical volume during one class term.

(iii) There shall not be more than nine instances of such multiple copying for one course during one class term.

[The limitations stated "ii" and "iii" above shall not apply to current news periodicals and newspapers and current news sections of other periodicals.]

III. *Prohibitions as to I and II Above*

Notwithstanding any of the above, the following shall be prohibited:

(A) Copying shall not be used to create or to replace or substitute for anthologies, compilations or collective works. Such replacement or substitution may occur whether copies of various works or excerpts therefrom are accumulated or reproduced and used separately.

(B) There shall be no copying of or from works intended to be "consumable" in the course of study or of teaching. These include workbooks, exercises, standardized tests and test booklets and answer sheets and like consumable material.

(C) Copying shall not:

(a) substitute for the purchase of books, publishers' reprints or periodicals;

(b) be directed by higher authority;

(c) be repeated with respect to the same item by the same teacher from term to term.

(D) No charge shall be made to the student beyond the actual cost of the photocopying.

Copyright Quiz

DIRECTIONS: Consider each of the following ten cases to determine which, if any, violate the copyright law as briefed above. Cover answers (below) until you have finished the quiz. What should be the standard for "passing" this quiz?

1 A teacher buys one consumable workbook and makes one copy of the workbook for each student in the class.

2 A teacher is preparing for a class presentation. The evening before the planned presentation, the teacher finds an article in a copyrighted journal that would be perfect for the class presentation. The teacher makes copies of the article for distribution to the entire class.

3 A professor provides the students of one class with copies of copyrighted articles on twelve different occasions within one semester.

4 A teacher makes a single copy of a chapter from a copyrighted book to be used as a reference for preparing a class lecture.

5 During one course, an English teacher provided the students with five short poems, copied from a book of poems, all written by the same author.

6 A student uses an opaque projector to enlarge a drawing so that it may be used to help illustrate an oral report.

7 A teacher makes a transparency, for the overhead projector, from a page of the class workbook.

8 A teacher makes a video tape of a rented 16mm film so that future students may see it.

9 A teacher makes a copy of an article for distribution to the class, but does not include the copyright notice on each copy of the article.

10 A student copies music from a record the student owns for background music for a class project.

Quiz Answers

1 Violation: See Guidelines III, B.

2 No Violation: See Guidelines II, Spontaneity(ii).

3 Violation: See Guidelines II, Cumulative Effect(iii).

4 No Violation: See Guidelines I, A.

5 Violation: See Guidelines II, Cumulative Effect(ii).

6 No Violation: See Guidelines I, D.

7 No Violation: See Guidelines I, D.

8 Violation: See Limitations on exclusive rights: Fair use. (4).

9 Violation: See Guidelines II, C.

10 No Violation: See Limitations on exclusive rights: Fair Use. (1), (4).

Circular R21

Reproduction of Copyrighted Works by Educators and Librarians

EXERCISE 21

DRAWING LETTERS

The Reason for It

Many instructional materials require the use of lettering; posters, flash cards, picture captions, titles for film and television productions, transparencies for overhead projection, and flip charts are a few of the projects in which hand lettering can be used for economical and effective displays. This exercise presents simple techniques for drawing letters by hand with felt pens. With practice, and by employing some helpful hints, you can develop the necessary skills.

Before You Start

Brown, Lewis, and Harcleroad: *AV Instruction*, Chaps. 5, 6. This manual: Exercises 8, 22; Correlated References, Section 30.

Purposes

1 To develop confidence in freehand drawing of letter with different types of felt pens

2 To learn to space letters for maximum legibility

3 To become acquainted with various types of felt pens that make different types of lines and effects

4 To present a standard lettering style—manuscript writing—which is used especially in primary grades, but may be a useful standard for other uses

Required Equipment and Materials

1 A variety of felt-tip pens

2 Newsprint or butcher paper sheets

3 Ruler or straight-edge

4 Pencils (HB or 2H grade)

Assignments

1 Study the illustrations below and on the following page. Note the basic techniques and actions suggested.

2 On a large sheet of paper, draw very light parallel lines; make several 1 inch apart, others 2, 3, 4, and 5 inches apart. Selecting from the several sizes and types of felt pens at your disposal, according to the spacing between the lines you have made, draw repetitions of the curves, circles, and lines that are common in all letters. See the illustrations. Use the technique of holding the pen firmly and move your arm to draw the form of the letter elements. When you have filled the lines of each height with the forms, stand back and evaluate your choice of pen tip for the height of your lines and circles. Decide where you could have made a better selection of felt pen.

3 Again, make light parallel lines—about 2 inches apart—on a large sheet of paper. Now, with a felt pen that you believe to be the best size and type, print a series of words that state an idea, a slogan, or a heading for a chart or display. Review your technique from time to time, and observe your letter spacing. What can you do to improve the lettering? Have you selected the best pen for your purpose?

4 Now, draw lines for uppercase and lowercase letters—that is, three parallel lines, one for the base of your letters, one for the top of your lowercase letters, and one for the top of your capital letters. Write a short sentence. Evaluate the results, and determine whether you need to practice basic lines, curves, and circles. Have you spaced your letters "optically" or "mechanically"? Redo the sentence, and improve your technique. Compare your hand and arm movement with motion illustrated in the picture on the next page.

5 After practice in the assignments above, prepare what you consider a finished display of lettering, at least three lines long; indicate in a small caption the purpose for which the lettering is designed: a flip chart for an audience of fifteen people; a title for a slide presentation. Have others evaluate your work.

6 By now you may feel the need to standardize your lettering. Try using the manuscript alphabet on the next page (and in larger format on the inside of the front cover) as your guide. Repeat assignment five using manuscript writing. Note the arrows on the sample that indicate directions of strokes.

Put some of your work on the bulletin board for evaluation. Ask: Are the letters readable?

After drawing parallel lines on paper, hold hand steady as you practice basic lines to be found in all lettering.

Helpful Hints

• Grasp your pen comfortably. As you letter, make sure you move your whole arm and not just your fingers. Letter at a moderate speed; if you move too slowly, you produce a jittery line, and if you move too rapidly, you will lose control and make irregular and poorly spaced letters. Keep the pen tip pointed toward the top of the page. Don't rotate the pen. By keeping the tip in a fixed position and with only arm movement, the lettering will have a pleasant "thick and thin" line.

• Be aware of the importance of optical spacing of the letters. Spaced mechanically, large letters are difficult to read, and have a loose appearance. See the sample on this page.

• Neat letters, hand-drawn, have a pleasant, informal appearance. Consider how the use of colors for your lettering, and the occasional use of symbols, can give sparkle to your work.

Completing the Assignment

Hold the fingers steady! Do *not* rotate fingers. Move only the arm in the various directions necessary to complete the letter.

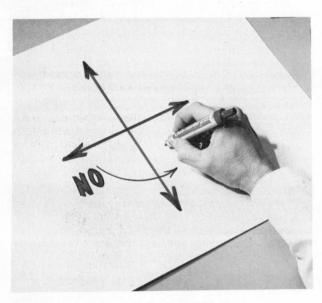

The tip of the felt pen should be pointed toward the top of the page for graceful "thick and thin" lines.

Space letters optically, not mechanically. Let your eyes tell you when the areas between letters appear to be equal.

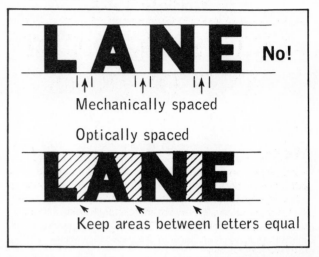

A variety of lettering styles may be produced by the many felt pens available.

In this drawing, stroke sequence and direction are indicated by numbered arrows. Remember to use optical spacing, keeping the areas between letters approximately equal.

Used with permission from Creative Growth with Handwriting, 2d ed., Zaner-Bloser, Inc., Columbus, 1979.

48

EXERCISE 22

LETTERING AIDS AND DEVICES

The Reason for It

When words are used to present ideas visually, legible and attractive lettering enhances communication. Aids such as lettering guides, stencils, rubber stamps, and a great variety of prepared letters are available to give a finished, professional appearance to lettered materials.

Before You Start

Brown, Lewis, and Harcleroad: *AV Instruction*, Chap 5. This manual: Exercises 21, 22; Correlated References, Section 29.

Purposes

1 To recognize the characteristics and applications of various kinds of lettering aids and devices
2 To gain experience using lettering aids and devices

Assignments

1 Examine the various types of lettering aids in your labora-tory. Practice using several of them. Select suitable ones, and develop materials for bulletin-board displays (Exercise 23), cloth-board displays (Exercise 28), additions to mounted materials (Exercise 32), transparencies (Exercises 35, 37, and 38), or for other instructional purposes.

2 Evaluate the lettering aids available. Complete the chart at the end of this exercise.

3 View the film *Lettering Instructional Materials* and pre-pare a statement that itemizes the principal uses of the various materials and techniques demonstrated.

Special Tips

• Select simple lettering styles that are easy to read.

• Use capital letters for short titles. But for five or more words, and for labels, use lowercase letters which are more easily read.

• For good separation, letter with a color that contrasts with the color of the background.

• Prepare lettering large enough for easy legibility for the purpose intended. For display materials use this guide:

Viewing Distance	Minimum Letter Size
8 ft	¼ in
16 ft	½ in
32 ft	1 in
64 ft	2 in

• Space letters by judgment—using "optical spacing." Give the effect of an equal area between letters rather than equal distance between them.

• When making a display that includes mounted pictures as well as lettering, always do the lettering before attaching the picture to cardboard backing; everyone can make a mistake in spelling now and then.

Cardboard cutouts.

Paper cutouts.

Gummed-back paper.

Plastic stencils.

Pressure-sensitive letters.

Cardboard stencils.

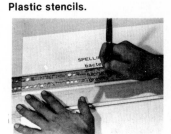

Lettering guides.

Mechanical devices.

Photocomposing machine.

WORKSHEET 22

Name _____

Course _____ Date _____

YOU MAY COPY THIS FORM TO COMPLETE THE EXERCISE

On the basis of the four criteria listed in the column headings below, evaluate each of the types of lettering you examined or used.

Types	Parts Required	Ease of Use	Quality of Lettering	Cost
Cardboard cutouts				
Paper cutouts				
Gummed-back paper cutouts				
Pressure-sensitive dry transfer letters				
Cardboard stencils				
Plastic stencils				
Lettering guides				
Mechanical devices				
Photocomposing machine				

EXERCISE 23

DESIGNING BULLETIN BOARDS

The Reason for It

Form and content are related. This generalization can be given specific meaning by considering layout as a skill. How you display words, symbols, and pictures; how you relate them in space; how you give them emphasis; and how you provide appeal to the mind through the eye may be keys to success or failure in communication. This exercise gives you opportunity to explore the possibilities of bulletin boards as important means of supporting and giving focus to instruction.

Before You Start

Brown, Lewis, and Harcleroad: *AV Instruction*, Chaps. 5, 6. This manual: Exercises 8, 21, 24; Correlated References, Section 5.

Purposes

1 To develop ability to plan displays

2 To practice organizing, planning, and executing bulletin boards

3 To understand how three-dimensional objects as well as flat surfaces may be used in bulletin-board displays

Required Materials

1 Bulletin board, either portable or wall

2 White paper or newsprint, about 8½ by 11 inches (21.5 by 28 cm)

3 Soft pencils (HB or 2B grade); soft rubber eraser

4 Construction paper, assorted colors; tracing paper

5 Scissors, rubber cement, and straight pins

6 Pictures of various sizes and shapes appropriate for the display theme selected

Beverly Hills Unified Schools, California.

Assignments

1 Select a title that represents a theme idea for a bulletin-board display. List ideas that help to illustrate the theme. Select pictures or other types of illustrations that can contribute to the theme or that represent the ideas you have listed.

Develop a shape for your display; study the shapes on the following page; recall others you have read about in the references. With a soft pencil, begin to sketch an arrangement for the items you have collected. You may find that arranging the illustrations on a table or on the floor will be helpful, sketching promising patterns as you evolve them. The frame line around your arrangement may be a square, a rectangle, or an informal shape; let your theme and materials guide you to the arrangement.

When you have drawn several "thumbnail sketches," with different forms and arrangements, select the one that most effectively communicates your theme. The sketch you finally select is your "layout design."

On the bulletin board or on a large sheet of paper, the proper size for your display materials and lettering, prepare your bulletin board for presentation and evaluation.

2 Select a theme such as "Early Earth Creatures" or "Nature's Wonders from Tide Pools." Design a bulletin board that includes one or more of the following elements: three-dimensional objects such as models, real things, small dioramas, or miniature sand tables. Captions and photographs may contribute effectively. Perhaps motion will add to the display; consider tiny electric motors, such as those used in store displays, and design for their contribution to the effect. Prepare your display and present it for evaluation.

Helpful Tips

• In developing thumbnail sketches, there may be a line or an idea in one sketch that you like especially, although the total sketch is not satisfactory. Use tracing paper to copy the portion of the sketch you want to use. Blacken the back of the tracing paper with pencil, and on another sheet of paper trace the lines you want to transfer. Do not use carbon paper; it will smudge.

• Check your layout for the following elements:

Balance. Are both sides of your layout in balance? Does one side appear heavier than the other? Push the parts of the layout around within your box or shape until you feel that both sides are about equal in weight.

Emphasis. Not all parts of a layout can be equal in importance. Decide which parts are most important; emphasize them with color, lines, borders, colored backgrounds, or by setting them out in front of the background with spacers.

Contrast. In order to be sure that important parts of your display can be seen, make sure you contrast light and dark elements. Again, backgrounds of color may help.

Harmony. Do the elements in your layout harmonize? Is the lettering of the style, weight, and size appropriate for the display? (See Exercises 1 and 2.) Do the sizes of illustrations, headings, captions, and the colors you have proposed make a pleasant, unified effect?

• For fastening parts of the display to the supporting surface, consider straight pins, rubber cement, double-faced gummed tape, transparent tape, white glue, and bulletin-board wax (a tacky, plastic substance that can be molded).

• Consider lettering captions on separate pieces of paper, color as desired, and mount them as part of the display design. All headings and captions should be included in the total layout design.

• Include in your bulletin board the means to involve those who will view it. Let the presentation develop an idea and have the viewers make deductions from what they see; provide a way for them to check what they have learned. For example, on the front of a piece of cardboard ask and question and mount the question flap on the display; under the flap, give the printed or typed answer. There are many variations of this technique including incorporating an electric board in your display with buttons to push that cause lights to indicate correct answers. (See Exercise 30.)

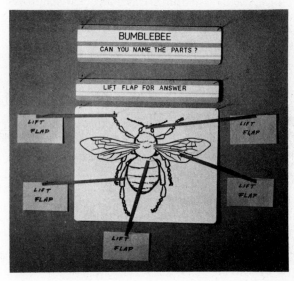

• If you have difficulty with a layout, these shapes may be an aid to planning.

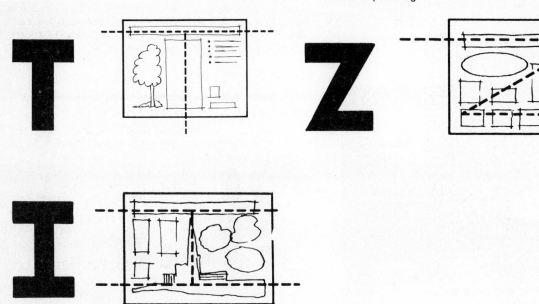

EXERCISE 24

DESIGNING DISPLAYS

The Reason for It

Simple, effective, three-dimensional displays about objects, people, and events can be important devices for communication. Presentations need not be restricted to wall surfaces; by locating items on tables or hanging them from the ceiling or from wall brackets, the entire room may be used for displaying.

Before You Start

Brown, Lewis, and Harcleroad: *AV Instruction*, Chaps. 5, 6; Reference Section 4. This manual: Exercises 8, 21, 23; Correlated References, Sections 5, 14, 40.

Purposes

1 To learn how to design a three-dimensional display

2 To practice making displays

Required Materials

1 Milk cartons, shoe boxes, round ice cream containers, and cardboard

2 Colored felt pens

3 Rubber cement, thumbtacks, and a stapler

4 Coat hangers and string

5 Scissors, razor blade, or sharp knife

6 Mounted pictures for backgrounds (see Exercises 31, 32)

Assignments

1 Prepare at least one three-dimensional display on any appropriate topic of your choice. Select your basic approach

A three-sided box for displaying flat pictures under a protective cover.

Carton shelf for simple diorama.

from one of the several display ideas illustrated in this exercise. Evaluate the effect.

2 Design a complete bulletin-board display that will use several three-dimensional objects. Consider having some objects on a table or shelf in front of the bulletin-board area. Prepare the final design for the display; include specifications for colors, lettering, and materials required. This assignment may be substituted for the one above.

3 (Project) Produce the display designed in 2, above, and present it for evaluation.

4 (Project) Develop and make a simple mobile display that can be hung from the ceiling or from wall brackets or lines stretched across the room.

Near the end of a course, some institutions may encourage preparation of such displays to be shown at an open house, along with other creative work done by students.

Planning the Display

1 As with any other instructional medium, you will want to plan before you begin work on the display.

2 Think of the objectives you wish to accomplish, and select or design a display which will make a contribution to the knowledge of the viewer.

3 Check reference material carefully to be sure that items you wish to include in your display are authentic.

4 For suggestions and ideas, examine pictures and sketches in resource materials of the event you wish to depict.

5 Make sure the figures you use are in scale with the scene.

6 Make a few thumbnail sketches. See Exercise 23 for tips.

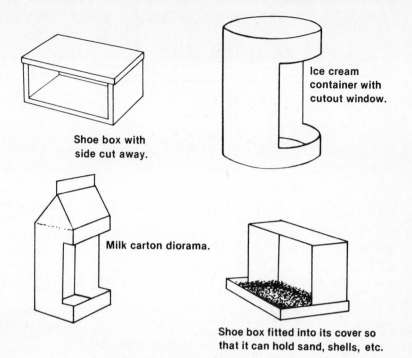

Shoe box with
side cut away.

Ice cream
container with
cutout window.

Milk carton diorama.

Shoe box fitted into its cover so
that it can hold sand, shells, etc.

Construction Ideas

Examples that accompany this exercise suggest a few simple techniques for constructing displays including shelves and clear plastic covered displays for flat pictures.

Ideas for Using Boxes

Pictures and sketches on this page suggest several ways to use boxes for displays. Given the opportunity, students will find it possible to develop practical variations to meet their special requirements.

Special Tips for Dioramas

• Small dolls or plastic figures add realism. When appropriate backgrounds are constructed in a diorama, stuffed animals or stuffed birds may be made to appear as if they are in their natural habitat.

• From readings, consider ideas that provide for audience participation, for using motion, and for including sound effects or projected images.

• Chicken wire covered with plaster of paris, sand, pine needles, and rocks helps to give texture to the floor of a diorama. A mirror can suggest a pond or a body of water.

• Boxes of various shapes and sizes can be adapted to form castles, houses, or furniture.

Ideas for Hanging Displays

Motion adds a dimension to suspended shapes, which may carry words, pictures, or other symbols.

Carton with painted background.

Coat hanger for "learning mobile"

Card stock panels

EXERCISE 25

SIMPLE SKETCHING

The Reason for It

Many people offer the unjustifiable excuse, "I am not an artist," as their reason for not making a drawing or a sketch that is needed to communicate ideas. Actually, simple cartoon drawing and sketching are within the ability of most teachers and students; interest, desire, and practice are the basic requirements. Developing bulletin boards or titles for a slide series or a television program, story telling, and making transparencies for overhead projection are only a few of the instructional projects that may be greatly enhanced by simple sketching techniques.

Before You Start

Brown, Lewis, and Harcleroad: *AV Instruction*, Chaps. 5, 6. This manual: Exercises 21, 22, 23, 24, 26; Correlated References, Section 37

Purposes

1 To discover how easy it is to sketch simple faces, figures, and scenes for instructional purposes
2 To gain experience in drawing faces and figures
3 To develop confidence in sketching subjects for visual communication

Required Materials

1 Sheets of newsprint or other paper
2 Felt pens, white and colored chalk
3 A hand mirror
4 Chalkboard or flip chart on easel

Assignments

1 Study the sketches on the page following. On a sheet of paper and using a felt pen, practice making the basic shapes for faces of men and women. Make shapes that seem to *you* to be interesting and that hold promise of being flexible in expression. Work without laboring as you make both front and side views; keep loose.

2 Having selected shapes, do a series of sketches with changing facial expressions. Make faces at your mirror to find the lines and forms that suggest emotions and attitudes. Note the effects of eye and mouth lines: create surprise, anger, sadness, puzzlement, delight, affection, and awe.

3 Develop on another sheet of paper a series of figures for the heads you have drawn. To what extent do you need to draw body positions and gestures to support the facial expressions? Note that the body of a cartoon figure may be only four to five heads high.

4 After completing male and female figures with postures and expressions that please you, copy several on a fresh sheet of paper. Now it is time to add identity, life, and warmth to your characters. To your figures add items of clothing, jewelry, decorations, or covering for the heads. Selected with care, these add-ons give identity to the sketches. And, as a final touch, use colored chalk to add to your drawings. Do not attempt to fill in faces and figures with color. With chalk of appropriate color, touch the edges of a line or a spot on the cheek, a few streaks in the hair, or a shadow area on the sketch; then rub and smudge the color for a soft effect.

5 Plan a few layout sketches that will represent a situation, a historical event, or a scene. Include a suggestion of background: Make simple, informal sketches to suggest buildings, city streets, airports, or other environments that can be suggested by a few lines.

6 When your practice is complete, arrange several of your sheets to show the evolution of your sketching from basic forms to finished sketch.

7 After you have established some characters and a style, in order to develop your skill with sketching when you are standing at a chalkboard or flip chart, plan and prepare a brief chalk-talk. Present your chalk-talk to the class, sketching as you go.

Simple cartoon faces can aid communication for both young and adult learners.

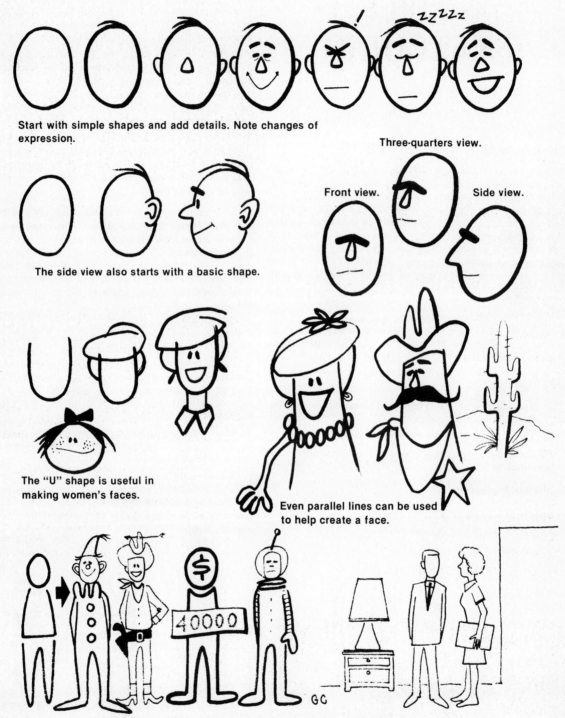

Start with simple shapes and add details. Note changes of expression.

Three-quarters view.

Front view.

Side view.

The side view also starts with a basic shape.

The "U" shape is useful in making women's faces.

Even parallel lines can be used to help create a face.

40000

GC

The body starts with three basic shapes. Add needed details and put in a background only if needed. Keep figures 4 to 5 heads tall.

EXERCISE 26

CHALKBOARD TECHNIQUES

The Reason for It

The least expensive, most readily available teaching tool—the chalkboard—is often considered only casually. As with any tool, some skill is needed to obtain from it maximum usefulness. Is the chalkboard to be a scramble of irregular, unplanned scrawls? Is it to be only a dusty record of yesterday's assignments? It can be an ever-changing center of dynamic communication, revealing creative skill and competence of both teacher and students.

Before You Start

Brown, Lewis, and Harcleroad: *AV Instruction*, Chap. 5. This manual: Exercises 21, 23, 24, 25; Correlated References, Section 6.

Purposes

1   To develop chalkboard lettering and writing skills
2   To learn how to transfer materials onto the chalkboard for instructional purposes

Required Materials

1   White and colored chalk (sharpened to get satisfactory line quality)
2   Card stock
3   Masonite or ¼-inch plyboard
4   Scissors
5   Window-shade material
6   Opaque projector
7   Motion picture projector
8   Filmstrip or 2- by-2-inch slide projector

Assignments

1   Analyze your present skill: Write a sentence on the chalkboard, and print a sentence below it. Ask other persons to read your work from various parts of the room, starting from the back; examine the work with them, listening for their comments. Can they read both sentences from all parts of the room? Are the lines level and straight? Are the chalk lines of sufficient thickness to be seen easily?

2   Draw several cartoon sketches and a diagram on the board, and include with them brief captions. Evaluate them as in the assignment above. What can you do to improve your skill and the effectiveness of your freehand work?

3   Prepare a simple template or a pounce sheet. For techniques, see the following page.

4   Produce a traceable image on the chalkboard using a projector—opaque 2- by 2-inch slide or overhead. You need not make an elaborate or finished drawing; complete enough of the image to evaluate the effectiveness of the size, techniques, and content you have used. Have others evaluate your work.

To Improve Your Work

• To produce a satisfactory quality of line, use a sharpened chalk.

• Work at the board with your elbow high, and walk along in front of the board as you write.

• To help keep the writing level, use dots on the board as starting and ending aiming points.

• In a conventional classroom, for legibility, writing or printing should be between 2 and 4 inches high; but for ensured satisfactory lettering and sketching on the chalkboard, choose lettering for each situation according to size of group, distance of viewers from the board, and type of content presented.

• When using colored chalk, use a soft chalk that can be erased easily. Always prime your board before using by dusting it with a coat of chalk dust; then smooth it with the eraser.

CHALKBOARD DRAWING AIDS

Template Method

When drawings, such as a state map, a basic structural form, or scientific equipment symbols, are often repeated on the chalkboard, templates simplify and speed the task.

1 On masonite or thin plywood, draw objects needed.

2 To cut out the template, use a coping saw or jig saw. Add a handle at the balance point of the template, if it is large or unwieldy.

3 The cutout template will provide a guide for rapid, accurate tracing.

Pounce Method

1 On window-shade material of appropriate size, trace or draw the image to be transferred to the chalkboard. (For occasional use, heavy paper may serve for your pounce pattern.)

2 To make a perforated line over your drawn lines, use a leather hole spacer.

3 With the pattern held firmly on the chalkboard, tap the surface over the punched holes with a well-chalked eraser. Chalk dust passing through the holes will provide a dotted guide line on the board that you can fill in with continuous lines either in advance or during your presentation.

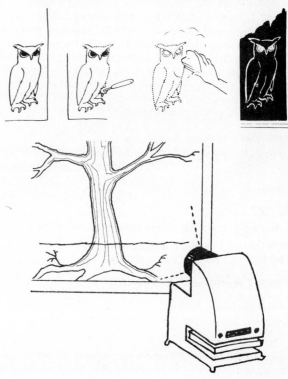

Projectors as Chalkboard Drawing Aids

1 Select a line drawing or scene from a printed page, a 2- by 2-inch slide, an overhead transparency.

2 Set up the appropriate projector needed to project the illustration you wish to transfer to the chalkboard. With room lights off or subdued, project and focus the image size that you require. Trace the lines necessary to present the illustration. By standing in front of the projector beam, you can evaluate the clarity and detail of your drawing. To clarify and emphasize elements in the drawing, use colored chalk when appropriate.

PLANNED CHALKBOARD WORK

Compare the two accompanying photographs. They suggest that planning the arrangement of a chalkboard display will improve its effectiveness and will facilitate student copying of drawn material. Refer to Exercises 3 and 4 on bulletin boards and displays for ideas on arranging chalkboard space for maximum communication through your presentations.

EXERCISE 27

GRAPHS AND TIME LINES

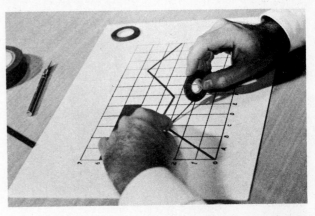

The Reason for It

To present the complex, interrelated information upon which many decisions are made today, graphs and time lines are commonly used. They save reader time and ensure accurate interpretation of information. Qualified teachers must be able to create graphs and time lines; students must be able to read and interpret them and, ultimately, to create them too.

Before You Start

Brown, Lewis, and Harcleroad: *AV Instruction*, Chaps. 6, 7. This manual: Exercises 21, 23, 24, 25, 26; Correlated References, Section 11.

Purposes

1 To learn some techniques involved in presenting statistical data and other kinds of information in graphic form

2 To learn to choose the correct type of graph or time line for presenting specific data

Required Materials and Equipment

1 Ruler, compass, and protractor

2 Colored pencils or colored chalk, lead pencil

3 Felt pens, assorted colors

4 Adhesive-backed drafting tapes, symbols, and colored and textured sheets

Assignments

1 If you are working alone, treat this assignment as totally your responsibility and select a graph project of your choice. Otherwise, work with a group and arrive at an agreement on a graph project for each student so that several types of graphs will be made. Proceed by discussing or considering basic kinds of graphs—circle, bar, and line, and time lines. Be sure that each participant understands the characteristics and advantages of each type of graph for presenting data. Examine the four examples of statistical information contained in this exercise. For each example, determine the type of graph that would best explain or present the data.

2 Have each student in the group select one of the projects for presentation or, if you are working alone, select one of the four projects. Design the graph that you believe will best serve to present the data; develop it as a lightly penciled sketch, using the layout tools mentioned above. When the rough sketch meets your requirements for accuracy and clarity, complete the drawing with felt pens. As required, add color with chalk, colored pencils, or adhesive-backed colors or textures.

3 If you employ a small sheet of paper for your graph, use an opaque projector to show it to your group or to the class for evaluation. Explain the information you have presented,

and obtain reactions to the type of graph and the techniques you have selected for your data. Does the graph communicate accurately, clearly, and quickly? If you have worked on a large sheet of paper to prepare the graph, display it on the bulletin board for evaluation.

4 As an optional presentation technique, you may prepare your graph in the form of a hand-made transparency, using appropriate materials.

Data for Assignments

1 Typical expenditures for X school for audiovisual equipment on a percentage basis were as follows:

Record players	20%
Tape recorders	25%
Filmstrip-slide projectors	15%
Opaque projector	10%
Motion picture projectors	30%

2 Temperature changes with elevation. Other things remaining the same, the higher one moves above sea level, the lower the temperature. The figures following are examples of such changes in one region:

Sea level	85 °F
1,000 ft	75 °F
2,000 ft	65 °F
3,000 ft	55 °F
4,000 ft	45 °F
5,000 ft	35 °F

3 Seating capacity for the meeting rooms in the Convention Center are as follows:

Plaza Room	175
Del Oro Room	150
Garden Room	100
Palmdale Room	50
Seneca Room	200
Patio Center	500

4 Given a list of the national parks in the United States and the date of establishment of each, present the data in graphic form.

A reusable line graph can be made by drawing lines on card stock and covering this with acetate on which graph lines may be made with grease pencil or water-soluble felt pen.

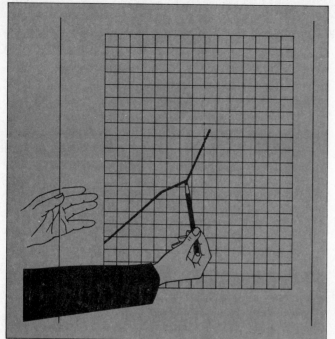

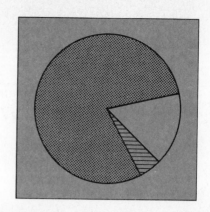

Circle or pie graph.

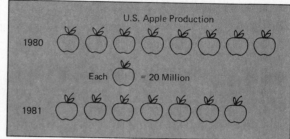

U.S. Apple Production

1980

Each ⬤ = 20 Million

1981

Pictogram.

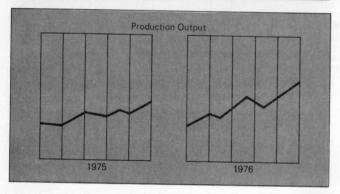

Production Output

1975   1976

Line graphs.

Bar graph—easily completed with colored tape—such as Chart-pak.

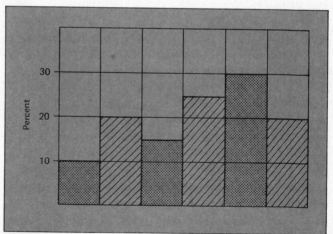

Time-line pictograph.

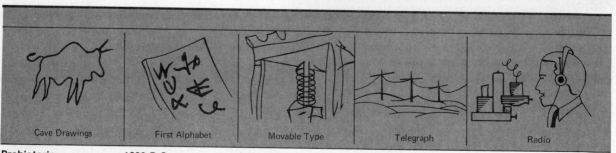

| Cave Drawings | First Alphabet | Movable Type | Telegraph | Radio |

Prehistoric.   **1600 B.C.**   **A.D. 1440**   **A.D. 1844**   **A.D. 1896**

EXERCISE 28

CLOTH BOARDS

The Reason for It

Cloth boards, inexpensive and easy to use, are limited in effectiveness only by teacher ingenuity. Flannel boards and nylon hook-and-loop* boards can help communicate abstract ideas, can be guides to organized thinking, and are effective tools for teaching, especially with small groups at any level of maturity.

Before You Start

Brown, Lewis, and Harcleroad: *AV Instruction*, Chap. 5. This manual: Exercises 21, 22, 25, 26, 27; Correlated References, Section 7.

Purpose

To provide guidance and experience in constructing and using flannel and hook-and-loop boards.

Required Materials

1 A prepared, ready-made flannel board or hook-and-loop board or materials, listed below, to make a cloth board

2 *Surface materials* to make a board: cotton flannel, felt, or hook-and-loop cloth

3 *Attachment materials:* sandpaper, pieces of flannel, or nylon loop material or pregummed loop strip material, or flock paper

4 *Bonding materials:* rubber cement, white glue, or staples

5 *Display materials:* pieces of flannel, lightweight cardboard, yarn, paper, or lightweight three-dimensional objects

6 *Backing materials* (on which to mount flannel or hook-and-loop material): Celotex building board, ½ inch thick of the desired size

Assignments

1 Obtain a prepared, ready-made cloth board, and design a presentation for a portion of a lesson, such as explaining a concept or a demonstration showing how some device operates. Construct the display materials in a form appropriate for the board you will use, applying the proper bonding material to your display items. Present the lesson segment to your class for evaluation.

2 Construct a cloth board of a material suitable for your need. Use hook-and-loop materials for a permanent and durable board with high holding power for display items. Either prepare a demonstration as described in 1, above, or make a series of permanent symbols suitable for your subject field that can later be used with different presentations.

* Many technological innovations derive their popular names from a single trade name; such is the case with the term "hook and loop," a name given to a nylon material that has found many applications. In the media field, the term is derived from cloth boards manufactured and sold under the registered trademark "Hook N' Loop" by Charles Mayer Studios, Inc., 168 East Market Street, Akron, Ohio 44308. For some applications, the nylon material is known as Velcro.

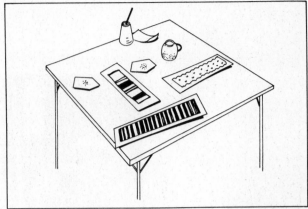

Charles Mayer Studios.

Making Hook-and-Loop Boards

One of the most durable and versatile materials for use as backing for display-board items is the nylon hook-and-loop fabric now available at yardage stores and audiovisual dealers. The clothlike loop material, which comes in a variety of colors, has a surface covered with tiny nylon loops. Strips of hook tape, surfaced with minute nylon hooks, are attached to the objects to be displayed. Tapes may be cemented to display materials; pressure-sensitive and solvent-activated tapes are also available. Advantages of hook-and-loop boards are: Supported materials do not slip or slide, attachment is firm; yet objects may be instantly removed or manipulated; both odd-shaped and heavy objects—up to several pounds—may be supported; and a wide variety of display ideas can be put into practical effect. Accessory materials include precut letters, small magnetic chalkboards, and pads of drawing paper backed by hook tapes.

Mount the nylon-loop fabric on a backing of Celotex, using a technique similar to that used for making a flannel board. However, because of the strong gripping action of the Velcro materials, overall attachment of the cloth to the board backing is desirable; use rubber cement spread *lightly* on both cloth and board surfaces and allow to dry before joining them (see Exercise 11 for rubber-cement mounting techniques). Edges can be finished as for a flannel board, or framed with light metal or wood.

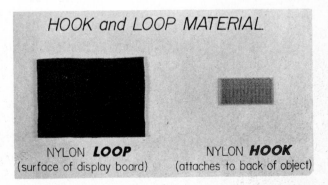

HOOK and LOOP MATERIAL

NYLON **LOOP** (surface of display board)

NYLON **HOOK** (attaches to back of object)

Preparing Materials for Hook-and-Loop Boards

Small patches of hook material will hold almost any substance to the Velcro board. Use rubber cement to attach patches to lightweight display items; use white glue or contact cement for heavier objects. Test the holding power of the hook-and-loop materials by experimenting with three-dimensional pieces of considerable weight.

ROLLER BEARINGS

Chales Mayer Studios.

Making Flannel Boards

Start with Celotex board, with the flannel or felt covering material, and thumbtacks or stapler. Fold the material over one edge of the board and thumbtack or staple on the back about every 6 inches. Turn the board over, stretch the covering slightly over the opposite edge, and fasten it in place. Finish the other two edges in the same way. For a smooth job, cover the edges and the loose ends of the cloth with wide masking or fabric tape.

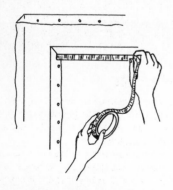

Preparing Materials for Flannel Boards

Many materials that have a nap or rough-textured surface will stick to a flannel board without special processing. Try such materials as colored yarns, steel wool, pipe cleaners, sandpaper, balsa wood, embroidery floss, and plastic foam.

Smooth cardboard, small three-dimensional objects, and other flat surfaces must be backed with rough texture or nap materials to give them holding quality for flannel-board use. A quick and inexpensive way to put rough texture on the back of display items is to staple on small strips of coarse sandpaper. Another inexpensive way is to use water glass and sand. Wipe or brush a liberal amount of water glass on the back of the cardboard and sprinkle coarse sand over the wet surface immediately. After a minute, when the water glass has set, lift the cardboard, and shake off the excess sand. The item is now ready to use.

Shapes, symbols, and designs can be cut from swatches of colored felt or flannel which will adhere well to the flannel-board surface. Use sketches or magazine pictures as outline templates for drawing the desired pattern on the felt. Felt, along with many other materials, also makes excellent background pieces for the flannel board—for example, sky, clouds, trees.

When illustrations from books, magazines, and free or inexpensive materials are badly worn, they may be converted to new uses on the flannel board by backing them with material that will adhere to the board.

Samples of commercially produced cloth-board materials.

EXERCISE 29

MAGNETIC BOARDS

The Reason for It

As with cloth boards, magnetic boards are especially useful with relatively small groups. However, some technical training schools and colleges, and an increasing number of lower schools, have large areas covered with light-colored, magnetic chalkboards. The surface of the board serves not only as a conventional chalkboard, but also as a projection screen area and as a surface for the manipulation of objects and graphic materials backed with magnets. The versatility of such a resource is great and promotes creative and interesting presentations.

Before You Start

Brown, Lewis, and Harcleroad: *AV Instruction*, Chap. 5. This manual: Exercises 21, 22, 25, 26, 27; Correlated References, Section 7.

Purposes

1 To motivate interest in making and using magnetic boards with effectiveness and skill

2 To stimulate the use of magnetic-board techniques in developing creative presentations

Required Materials

1 A ready-made magnetic chalkboard, or materials to construct one; and a variety of sizes of small magnets

2 Attachment materials: masking tape, cement, or glue for attaching magnets to display materials

3 Display materials: lettering, graphics, and pictures mounted on cardboard, or three-dimensional objects to be displayed

REPARTITION DES ZONES DE VENTES

4 Materials for constructing a magnetic board: an oil drip pan, obtainable from an auto supply store; spray-on or paint-on chalkboard paint (color as desired); chalkboard chalk, and an eraser

Assignments

1 If necessary, construct a magnetic board.

2 Prepare a magnetic-board presentation on a subject of your choice. Include the following techniques to demonstrate your competence and creativity: Use mounted pictures, some real things, and lettered captions, using magnets to attach them to the surface of the magnetic board. Also, use white or colored chalk to add information, symbols, or connecting lines. When you select your topic, think of opportunities to manipulate the materials, to place them in position in sequence, or to remove them. Present a clear, interesting explanation as you use the board.

3 Plan and prepare a presentation that demonstrates how a magnetic chalkboard can be used as a screen for projected still images that are supplemented by relevant magnet-backed objects or pictorial or printed material. After your presentation, invite evaluative comments and suggestions from class members.

Making Magnetic Boards

Inexpensive magnetic boards can be made from oil drip pans available from hardware or auto supply stores. These pans have rolled edges and are usually sufficiently rigid so that further backing is unnecessary.

A magnetic board can also be made from a sheet of light-weight galvanized metal, available at hardware stores or sheet-metal shops. If the metal is thin, mount it with contact cement on a sheet of fiberboard, such as Celotex, or light plywood. Cover sharp edges with molding or heavy protective tape.

Some teachers have combined the advantages of magnetic boards with cloth boards (Exercise 8) in this way: The back of an oil drip pan is prepared with chalkboard paint, the front, indented surface is left unpainted. A sheet of fiberboard the size of the bottom area of the pan is covered with hook-and-loop (or flannel) material, as described in Exercise 8. This sheet is then cemented to the indented area of the pan. Thus, a double purpose, portable board is always available.

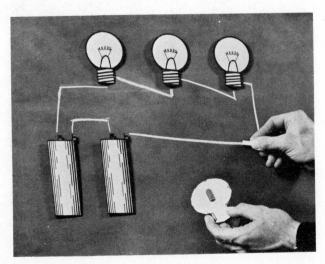

63

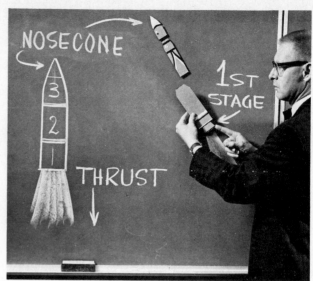

To prepare the board for use, first clean the metal with vinegar, and then wash it thoroughly with water. After the surfaces are dry, spray or brush on several coats of chalkboard paint. A very light sanding of the paint between coats is desirable; use fine sandpaper or steel wool. Finally, thoroughly chalk surfaces to ensure smooth writing and easy erasure of chalk marks. Be sure the board is completely dry before chalking, and do not write on the board until it is completely chalked.

Preparing Materials for Magnetic Boards

Inexpensive small magnets can be purchased in quantity. A size approximately ⅛ by ¼ by ¾ inches will serve many bulletin-board purposes, though larger magnets may be needed for some displays. Rubber magnets in strip form are also available and are convenient for some applications. Magnets glued or taped to the back of the item cause it to stand out slightly from the surface of the board, giving additional emphasis to the display. Strong magnets can also be used on top of sheets of paper to hold the paper on the board surface.

Tape one magnet near each end of a strip of cardboard. Try using it on the magnetic board. Insert magnets into some small objects, and use these on the board. Be aware of how these can be moved from one area of the board to another without having to remove them from the surface of the board.

Practical Pointers

To attach magnets to display materials, tape, rather than glue or cement, is recommended; the magnets can be removed easily from the tape, and the tape tends to protect the surface of the magnetic chalkboard from becoming scratched.

For some types of subject matter, frequency of repeated use justifies painting or otherwise applying lines or designs on the surface of the board. For occasional use, the back of a steel filing cabinet can serve as a magnetic chalkboard, another indication that ingenuity can lead to the discovery of useful resources, which are often close at hand.

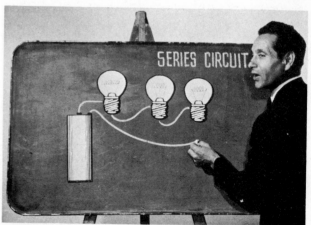

EXERCISE 30

ELECTRIC BOARDS

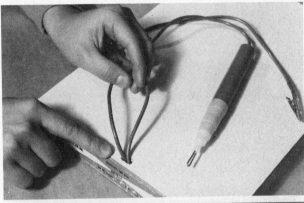

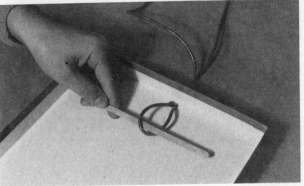

The Reason for It

As a means for rapid self-testing, the electric board gives students an active and interesting procedure. Such boards also provide opportunity for exploration of principles and for drill in mathematics and other skills. Electric boards can be used at all levels, and may be located in classrooms, learning centers, and in carrels. The method used to make the simple test board described here may be used also to make electric posters, bulletin boards, and charts. Involve students in electric-board construction, and they learn as they plan and make them.

Before You Start

Brown, Lewis, and Harcleroad: *AV Instruction*, Chap. 2. This manual: Correlated References, Section 8.

Purposes

1   To design and construct a box base for an electric board
2   To design and prepare electric-board circuit cards

Required Equipment and Materials

1   Self-powered electric circuit tester, obtainable at local hardware and electronic stores or by mail from PLA, P.O. Box 9726, San Jose, Ca 95106.

2   Scissors, paper cutter, craft knife; tongue depressor or stick; rubber cement or all-purpose glue

3   Light cardboard, manila folders, tagboard, or heavy construction paper; typing paper, notebook paper; butcher paper

4   Pencil, ruler; masking, cellophane, or package-sealing tape

5   Hand-operated paper punch that makes holes ¼ inch or larger

6   Aluminum kitchen foil

7   Cardboard box about 10 by 12 inches (about 25 by 30.5 cm) or larger (the depth is not critical; a typing paper box works well)

Assignments

1   *Develop at least eight matching test items* in your subject field and level. (For example: Q. What is the formula for water? A. H$_2$O.) Validate your answers by consultation or references, and test the questions with a group of typical students.

2   *Construct a box for an electric board,* as follows:

Punch a hole with scissors or a knife near the bottom of the lid of the box you have selected.

Fold the circuit tester cord near the middle, and insert it through the hole; turn the lid over, and slip a stick or tongue depressor through the cord loop, and tape it in place. Put the lid back on the box. The box portion of the electric board is now complete.

3   *Construct an electric board circuit card.*

Cut a piece of cardboard, or similar material, to 8½ by 11 inches (21.5 by 28 cm) in dimensions. Draw dividing line down the middle. Leave a space at the top for a heading. Below the heading space, draw horizontal lines dividing the card into eight pairs of rectangles. Refer to the pictures to see the arrangement and some satisfactory alternatives.

Punch holes with the paper punch on the left-hand and right-hand sides of the card, one hole for each rectangle. Set aside.

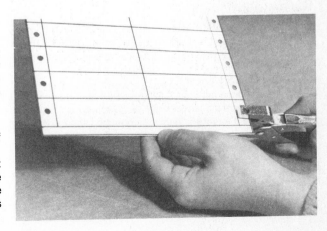

With rubber cement, coat a sheet of butcher paper about 8 by 14 inches (20.5 by 34 cm) in dimensions. Press a sheet of kitchen foil onto the cement-coated surface. Cut the foil-covered paper into strips about ⅓ inch wide and the length of the paper. These are circuit strips and will take the place of wire in the circuitry.

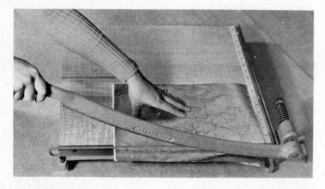

Place the circuit card *face down*. Tape a circuit strip, foil side down, with one end over a hole on the left-hand side and the other end of the strip over a hole on the right-hand side. Let your cross connections between the holes be in random sequence—that is, crossing the sheet diagonally in random order. When you have taped down the foil circuit strips for all the holes, trim off the excess lengths.

Coat the back of the circuit card, covering the criss-crossed strips with rubber cement. Overlay a sheet of paper and adhere it to the card; this sheet protects the attached foil circuit strips, and hides the direction of the interconnections.

Turn the card over. Print or draw the program on the surface of the card, questions on the left, matching answers on the right; be sure the cross-connection is correct.

Lay the circuit card on the question board. When the probe of the circuit tester is touched to the foil in the hole beside the question and the alligator clip touches the foil next to the correct answer, the lamp in the tester will light.

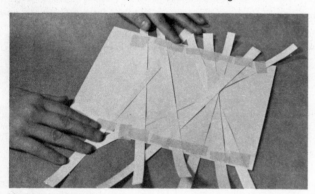

Helpful Hints

To reduce the possibility of students memorizing connection patterns, cards may be made with a variety of circuit arrangements. Circuit programs may be color-coded with matching question-answer sheets.

Question-answer sheets may be prepared in a number of ways: Pre-gummed labels, available in stationery stores, can be used to prepare questions and answers on a typewriter and applied to the rectangles on the punched sheets. Boards may be made in sizes required to accommodate photographic prints, sketches, or small objects taped or cemented to the question rectangles.

Consider preparing some question-answer electric-board connections on a large poster or display. Long strips of foil can serve for the interconnections between identified points and the question boxes in a corner of the display.

MOUNTING PICTURES

The Reason for It

In classroom activities and projects, many types of pictures and graphic materials are used. Teachers should be acquainted with inexpensive, simple, and quick methods of mounting such materials for effectiveness and ease of use, for attractive display purposes, and for efficient storage.

Before You Start

Brown, Lewis, and Harcleroad: *AV Instruction*, Chap. 9. This manual: Exercises 24, 36; Correlated References, Section 38.

Purposes

1 To gain experience in mounting pictures by the *dry-mount* method, using either a hand iron or a mounting press

2 To learn to use the rubber-cement method of dry mounting

DRY-MOUNT METHOD
Required Materials and Equipment

1 Two magazine pictures (tear sheets)

2 Two sheets of dry-mount tissue*

3 Two pieces of mounting board (6- to 14-ply cardboard)

4 An electric hand iron

5 A dry-mount press and tacking iron

Assignment

Study the two sets of pictures on the right. Practice each method by making at least one mount. Be prepared to discuss the effectiveness of the methods.

Do You Know?

Dry-mount tissue is a form of tissue paper coated on both sides with an adhesive material that becomes sticky when heat is applied. Set a hand iron at "low" or "rayon" and a mounting press at 225°F; set a tacking iron at *medium*. By placing the tissue between the picture and the mounting board and applying heat and pressure, a permanent attachment is made.

* With the hand iron use Seal product, Fotoflat. For the dry-mount press, use a regular dry-mount tissue, such as MT-5.

USING HAND IRON **USING PRESS**

After drying picture with iron, or in press, tack tissue to back of picture.

Trim picture and tissue together.

Tack tissue to mounting board (two opposite corners).

 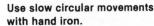 

Use slow circular movements with hand iron.   Apply heat and pressure (10 to 15 seconds) in press.

Protect and display completed mounts.

RUBBER-CEMENT METHOD
Required Materials

1 Two tear sheets
2 A jar of rubber cement
3 Two pieces of mounting board
4 Two sheets of household waxed paper

Assignments

Since mounting pictures with rubber cement requires no special equipment, undertake to use both the *temporary* mounting technique and the *permanent* technique.

1 Follow, step by step, the permanent method of mounting pictures illustrated on this page. Note especially the use of waxed-paper sheets.

2 After completing the permanent mount, mount another picture using the temporary technique. When the cement dries, remove the print with care. Occasionally there is need to remove photographs, thus the importance of the temporary mount procedure.

Do You Know?

If rubber cement is spread on the back of a picture and the picture is then set on a piece of cardboard (before the cement dries), a *temporary* mount will result; that is, the picture can be removed. If the cement is brushed on *both* the back of the picture and on the mounting surface, permitted to dry, and then the surfaces adhered, a *permanent* mount will result.

Review Questions

Reviewing your experience in mounting by both the dry-mount and the rubber-cement methods, discuss the following questions as a class, or in small groups, and share your judgments.

1 Which method of dry mounting with tissue is faster? Does the faster method also result in a more satisfactory mount?

2 Considering time, materials, costs, final results, and probable permanency, evaluate the efficiency of the three methods used for mounting (hand iron, dry-mount press, rubber cement).

3 When using dry-mount tissue, why is a picture first left untrimmmed, but trimmed to the desired size before using rubber cement?

4 From a storage standpoint, what is an important consideration when mounting a large number of pictures?

5 What values for teaching do you see in one or more of these mounting methods?

6 What methods can be used to mount maps or large-sized pictures on cloth in order to roll or fold them for long life and easy storage? (Check the references on the production of materials for suggestions.)

1 **Trim the picture to size.**

2 **Mark the placement of the picture on the mounting board.**

3 **Apply cement to both the back of the picture and the mounting board.**

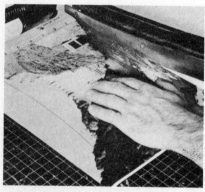 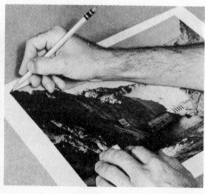 

4 **When the cement has dried, cover the board with wax-paper sheets. Place the picture on top of the wax sheets and slip the sheets out (one at a time) to each side. Rub the picture for good adhesion.**

5 **Remove excess cement from the board around the picture, as shown, or use a rubber cement "pickup" made from a ball of dried rubber cement.**

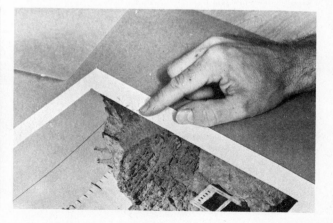

EXERCISE 32

LAMINATING

The Reason for It

When media are supplied for learning resources they must be durable, attractive, and effective. Among the techniques that support these qualities is laminating—the covering of flat resource materials with protective plastic sheeting. By lamination, mounted pictures, specimens, charts, game pieces, documents, and displays remain clean and durable for use, storage, and reuse. This exercise is to ensure that you know about the techniques and values of laminating.

Before You Start

Brown, Lewis, and Harcleroad: *AV Instruction*, Chap. 9. This manual: Correlated References, Section 40.

Purposes

1 To explore applications of laminating materials

2 To gain experience in applying *cold* (Con-Tact) and *hot* (Seal-lamin) protective plastic sheeting to flat displays

COLD LAMINATING METHOD

Required Materials and Equipment

1 A mounted picture, chart, or other flat display

2 A sheet of pressure-sensitive, clear plastic sheet (Con-Tact) at least 2 inches longer and wider than the display dimension

3 A 12-inch ruler; a pair of scissors; a large spoon or other smooth, hard-surfaced instrument for burnishing

4 A clean, hard, flat working surface

Assignment

Laminate a mounted visual, using cold pressure-adhesive plastic film, usually Con-Tact. This tough, thin, transparent plastic sheet is available in bulk rolls of different widths. A paper backing, marked to facilitate cutting, must be removed to expose the adhesive surface for application. When the sheet is pressed on a flat display, or over thin objects mounted on cardboard, they are securely protected. Follow these steps:

1 Cut the Con-Tact sheet to a size about 2 inches wider and 2 inches longer than the display to be covered

2 Peel the backing from one corner of the Con-Tact plastic surface to expose a 4- or 5-inch triangle of adhesive. Fold and crease the backing corner flat as shown. Do not touch the adhesive to any surface until ready to bond.

3 With the mounted visual face up and the backing paper on it, adjust placement of an equal border of Con-Tact around the visual. Remember, do not let the exposed adhesive touch the visual.

4 Lower the exposed adhesive corner to bond first with the visual, then with the table top. Avoid trapping air or allowing wrinkles to form.

5 Reach under the Con-Tact sheet, grasp the peeled corner of the backing sheet and pull it toward the opposite corner; at the same time draw the edge of the ruler along after the line of contact to smooth the plastic as the adhesive is exposed.

6 Trim Con-Tact at 45 degrees on each corner as shown. Fold the 1-inch border of the adhesive around to the back of the mounting board, and press it into place. This protects the edge of the board.

TRIM

7 Rub out any bubbles and thoroughly adhere the plastic to the surface, using fingers, a spoon, or other suitable burnishing tool. Persistent bubbles can often be punctured with a pin and rubbed flat. Work trapped air toward the nearest edge.

Notes and Cautions

• To laminate unmounted sheets on both sides, proceed to do steps 1 through 5; then trim the plastic to register with the visual and rub for a perfect bond; then repeat the steps on the opposite side.

• If leaf and flower specimens are to be laminated, they should be dried between sheets of absorbent paper under a weight for two weeks. A hot press may be used for drying, but there is a risk of losing true colors. Fasten dry specimens lightly in place on the mounting board with rubber cement before laminating. Shave thick stems thin on the back side, or cut channels for them in the board; sometimes blooms and foliage are composed on a mount with a drawing of the stem instead of the real object.

• For this cold method, do not use heat during or after applying the laminate; heat tends to cause draw, reticulation, and other distortions of the plastic and may alter the behavior of the adhesive.

HOT LAMINATING METHOD

Required Materials and Equipment

1 Sheets of Seal-lamin laminating material

2 Dry-mount press and tacking iron

3 Clean newsprint or Seal-release carrier

4 Seal-weight or magazines for cooling pressure

5 Scissors and ruler

Assignment

Laminate a mounted flat display, coating only one side of a mount. The laminating material suggested here is one of several satisfactory brands—Seal-lamin. It is a thin, tough, transparent mylar plastic available in rolls of various widths or in the form of envelopes. One side is coated with thermoplastic that melts at 275 to 325 °F, depending upon the object being laminated. When cooled below 270 °F, the thermoplastic congeals and bonds with an adjacent surface. Follow the steps below, and observe the Notes and Cautions that follow.

1 Set the hot press for 270 to 325 °F according to the thickness of the piece to be laminated; also see the manufacturer's recommendations included in the film package. Allow time for the press to heat.

2 Trim a piece of Seal-lamin film to cover and extend 1 inch beyond each edge of the mounting board. Flatten the laminate by stroking with hand or pad of lint-free cloth. If curl persists, tape the sheet out flat. The dull side of the laminate should be up and the display image placed face down on it. Center the display on the film to give equal 1-inch margins on each side. Trim the laminate diagonally to the corner of the mounting board, and trim as shown. Avoid dirt or lint on any surfaces.

3 Fold one margin of the laminate to the back of the mount, and tack in place with the tacking iron set on "high." Avoid slack or tension wrinkles in the film. Use a thickness of paper between the iron and the film. Repeat the process on the opposite edge, then the last two edges.

4 Sandwich the mount and laminate between the sheets of clean paper, image face up, and press at the proper temperature for 1 minute. Remove the mount quickly, and place it under a stack of magazines or a Seal-weight for a few moments. If bubbles or other signs of air trapped under the film appear, perforate them and press the mount again for 1½ minutes.

Notes and Cautions

• *Important:* some photographic print paper emulsions and some plastic or coated-surface magazine papers are destroyed by the temperatures required in this heat laminating process. Some printing ink pigments can also be damaged. Run a test sample of your material before risking damage.

• More pressure by the press is required for laminating than for dry mounting or for applying cloth backing materials. Seal, Inc., recommends using a ¼-inch sheet of masonite under the foam pad of their press for laminating work. But do not use any material thicker than ¼ inch; the press can be damaged.

• Since some mounting boards may curl, minimize this problem by reheating and holding the board in a reverse curl as it cools. Another technique is to fold the laminating film around the entire mounting and laminate both sides at the same time, pressing for 1 minute at 325 °F. The latter technique is used also when preserving images on both sides of a mount.

• If the mounting is wider than the available laminating film, laminate in sections and butt or overlap the laminate strips. The joints will be barely noticeable.

• Some graphic services that do volume work use motor-driven roller machines for doing both hot or cold laminating, on either one or both sides.

• If grease pencils or washable ink, fiber-tipped pens are used, the laminated surfaces can be marked and markings erased.

• Creative effects can be made with the Seal-lamin laminating process. For example, shave colored wax crayon on a mounting board, add some glitter, and laminate with crinkled laminating film. Try the same technique using random shapes and sizes of crinkled laminate applied in layers as on a collage. This technique can also be effective for overhead projection; mount the laminates on clear plastic transparency film; project, using a polarizing filter-spinner. Effects of age and textured surfaces can be created by using crinkled laminating film.

• When laminating is in demand, a laminating machine, below, conserves time and material.

EXERCISE 33

SPIRIT DUPLICATING

The Reason for It

The spirit duplicator is simple to operate. It is an efficient, economical tool for producing multiple copies of many different types of instructional materials. Multicolored spirit carbons and thermographic masters add still further variety and flexibility to the process. All teachers, and all but the youngest students, should be able to use spirit duplicators.

Before You Start

Brown, Lewis, and Harcleroad: *AV Instruction*, Reference Section 2. This manual: Correlated References, Section 15.

Purposes

1 To help you develop skill in preparing, laying out, typing, or drawing materials for spirit duplicating

2 To provide basic information about operating the spirit duplicator

3 To explore uses for duplicated teaching materials

Required Materials and Equipment

1 Drum type of spirit duplicator

2 Typewriter, electric preferred

3 Ruler, lead pencil, ball-point pen

4 Single-edge safety razor blade or X-acto knife

5 Wax correction pencil or cellophane tape

6 A commercially prepared spirit master pack or a duplicating carbon and a sheet of good-quality master paper; a second sheet of duplicating carbon in a contrasting color

Assignments

Read materials contained in this exercise. Then prepare a *single-page* spirit master on which you perform the following processes:

1 Type one paragraph of at least three lines with your typewriter ribbon set on "black."

2 Type a second paragraph of the same words with the ribbon set at "stencil."

3 Using a ball-point pen, draw the outline of a magazine or newspaper picture, following the procedure described in "Drawing on Masters," in this exercise.

4 Using the same ball-point pen, draw the same outline picture by simply removing the tissue and drawing directly on the face of the master sheet.

5 Type the words "and they" and "whenever he" as they appear in the exercise section, Making Corrections, on the next page. Then make the corrections that are illustrated. For the "cover up" correction, you may use small pieces of cellophane tape.

6 Using a standard spirit carbon paper, such as purple or black, type this paragraph on the master:

For the *second step*, carefully clean the lens surface with a camel's-hair brush. Do not fog the lens with your breath.

7 Directly below the paragraph just typed for Assignment 6, type this following paragraph using a spirit carbon of contrasting color, such as red or green:

For the *first step*, use a syringe or lens brush with air bulb to blow off any dust that can be removed by a blast of air.

8 Then cut the master to enable you to put the *first-step* paragraph ahead of the other. Be sure to cut evenly, preferably with a paper cutter. Hold parts together by applying cellophane tape at joints on the front of the master. Carefully remove or cover any carbon lines that may appear at the line of cutting.

9 After reading directions under Running Copies, run ten copies of your master. Be sure to balance the top and bottom margins and adjust pressure, as needed, to produce copies of acceptable density.

10 Study results. Enter notes on the margin of one of your copies to indicate: (1) which specimen, 1 or 2, you prefer, and why; (b) which drawing, 3 or 4, you prefer, and why; (c) which of the three correction methods used in 5 you found easiest or most satisfactory. Did you produce a neat splice on your master? Are your copies balanced and bright? Are you satisfied with the color contrast of the two carbons?

The Master Pack

The master pack consists of a sheet of master paper and a sheet of spirit carbon paper; carbon side is up, facing the master paper. When an impression is made on the master paper by typing or drawing, carbon is transferred to the back of the master.

Typing on Masters

1 *Prepare the typewriter.* Clean the keys thoroughly, using a stiff brush or key putty.

2 *Plan your layout.* Visualize or sketch how your final copy will be arranged.

3 *Insert the master pack into the typewriter.* As you see it in your typewriter, it should be arranged in the order shown here with the carboned side of the carbon paper toward you.

Using Color

Standard spirit-duplication-process carbons are purple or black. But you can make printed impressions in other colors by using red, green, or blue carbons. An advantage of the spirit duplicating process is that you may add one or more of these colors to your master by substituting color carbons at points where you wish to write, type, or draw. All colors print simultaneously on a single run of the duplicator.

Making Corrections

If while typing a spirit master you make an error, you simply have carbon where you do not want it. To correct a mistake:

1 *If you do not need to type over the mistake,* cut it out with a razor blade, a sharp knife, or a small pair of scissors. Or cover the error with transparent tape, a piece of gummed label, or a special wax pencil supplied for this purpose. Either method is good for eliminating an extra letter between words, a word that runs into the margin, or an unwanted underscore or punctuation mark.

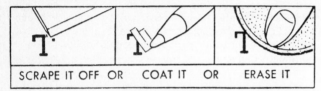

and|t|they | and|t|they | and|they
whenever |he| | whenever |he| | whenever
CUT IT OUT or | COVER IT UP or | BLOCK IT OUT

2 *If you must make a correction at the point of error,* you must first eliminate the error. Do this by (1) lightly scraping off the carbon with a razor blade, (2) coating the error with wax from a special pencil, (3) erasing the error thoroughly, or (4) scraping and wax-coating the error.

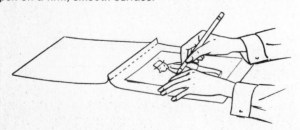

SCRAPE IT OFF OR | COAT IT OR | ERASE IT

3 *After the error has been eliminated,* type the correction. Remember that you have already used up the carbon at the point of the error, so you need new carbon there. Insert a slip of fresh carbon at the point where retyping is needed. Remove the slip before going on.

4 *Use helpful shortcuts.* Scrape parts of letters with a razor blade to produce other letters—e.g., changing an E to F, and o to c. To remove entire sections or to rearrange the order of copy, cut masters in two or more pieces. Splice the sections together by applying cellophane tape to the front surface of the master.

Drawing on Masters

Practice transferring drawings or diagrams to spirit masters by these procedures:

1 *Freehand drawing and lettering.* To draw directly on the master sheet, fold back the carbon sheet from the master sheet, as shown. Sketch your drawing or lettering lightly in pencil on the master sheet. Fold the carbon back under the master sheet, leaving the slip sheet out. Go over the drawing or lettering on the master, using a sharp pencil or a ball-point pen on a firm, smooth surface.

2 *Tracing.* To transfer illustrations or lettering from magazines, clip art, or previously prepared drawings, attach the material to the master sheet with tape. Remove the slip sheet from the master. Trace over the lines to be reproduced with a pencil or ball-point pen, using firm, even pressure. The image will appear in mirror fashion on the back of the master sheet. Tone areas by shading with dots or cross-hatchings. Add interest to drawings by applying solid color in small areas or by outlining with lines of different weights.

Running Copies

Read the instruction book accompanying the duplicating machine you are to use. The following are typical general procedures:

1 *Prepare the duplicator.* Be sure there is fluid in the tank, that the paper feed and grippers are adjusted, that the pressure level (if applicable) is set at "medium," and that the wick is moist.

2 *Ready the master.* If you use a ready-prepared master set, remove the separating tissue and tear the master from the carbon back.

3 *Attach the master.* Turn the drum so that the master clamp is up. Open the drum clamp, insert the master (with carbon side up, toward you), and close the clamp.

4 *Run the copies.* Turn on the duplicator motor or turn the machine handle. With some machines it will be necessary to turn the handle (counterclockwise) first to the "four o'clock" position, and then clockwise to make the first copy.

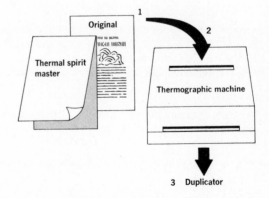

Making Thermal Spirit Masters

To transfer drawings as well as typed copy to a thermal master:

1 Place material to be copied under the thermal spirit master, as shown. Remove the slip sheet.

2 Insert this "sandwich" into a thermographic machine (Thermofax or other). Machine heat causes carbon dyes to transfer to the master.

3 Carefully peel master from the carbon.

4 Attach master to duplicator and run in usual fashion.

EXERCISE 34

ELECTROSTATIC DUPLICATING

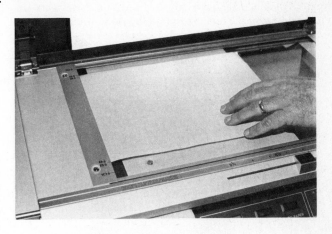

The Reason for It

The electrostatic copying process has many useful applications for educators. It provides quick, legible copies from prepared originals on paper or acetate. Some copiers have the ability to reduce the size of copy images; others can reproduce in color. Electrostatic copying produces dry, direct-reading copies from pencil, typed, carbon, or colored ink originals.

Before You Start

Brown, Lewis, and Harcleroad: *AV Instruction*, Reference Section 2. This manual: Correlated References, Section 15.

Purposes

1 To help you learn how to create an original master for electrostatic reproduction

2 To develop your familiarity with basic tools and materials for creating an original master for electrostatic duplicating

Required Equipment, Tools, and Supplies

1 Access to an electrostatic copier

2 A 12-inch ruler

3 A single-edge razor blade

4 A flat, smooth work surface (table or drawing board)

5 A blue pencil

6 Black ink and pen

7 Adhesive (such as rubber cement)

8 A mounting board (such as white poster board or white paper)

9 Prepared art (to be duplicated)

10 Masking tape

11 White opaquing fluid (to correct errors, smudges)

Assignment

Design and prepare an original master for electrostatic duplication. Use the following: black ink line; prepared art; typed or lettered copy; and prepared lettering. Follow directions presented in Procedures.

PROCEDURES

Follow directions accompanying the sketches below to design and produce an original master for reproduction by the electrostatic process.

1 Tape the mounting board to the table or other work surface. Use masking tape for this purpose.

3 Produce the original master. Include in your master examples of: black ink lines, prepared art, typed copy, hand-lettered copy, and prepared lettering. Use rubber cement or glue stick to attach prepared art. Rub off any excess rubber cement. Use a ruler to align the lettering or inked lines.

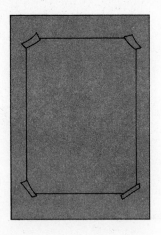

2 Use a blue pencil to lightly draw elements of your layout. Remember that blue pencil will not reproduce, if lightly applied. Erase carefully.

4 Make one clean electrostatic copy of your original master. Use white opaquing fluid to "white out" any errors or smudges on the copy. Sign your name on the back of your sample copy and turn it in to the instructor.

5 Keep your original master for future copying requirements. It may be changed and updated, as required.

Adapted from materials by Leonard Espinosa, San Jose State University.

EXERCISE 35

HANDMADE TRANSPARENCIES

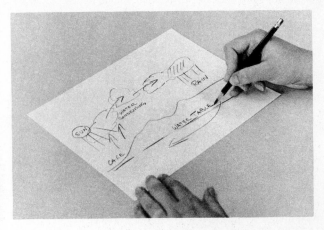

The Reason for It

The extensive use made of overhead projection techniques by both teachers and students at all levels of education has stimulated projection of handmade transparencies. With a few simple materials and minimal equipment, effective transparencies can be made quickly and economically. Also easily mastered are the techniques for making overlays, adding color, and even making parts that move or using special materials to simulate motion when movement will clarify the presentation.

Before You Start

Brown, Lewis, and Harcleroad: *AV Instruction*, Chap. 7; Reference Section 1. This manual: Exercises 36, 37, 38; Correlated References, Section 15.

Purposes

1 To investigate ways to prepare low-cost, easily produced transparencies for use on the overhead projector

2 To gain experience in planning and making transparencies of the following types:

   a Hand-drawn, with overlays

   b Motion transparencies

HAND-DRAWN TRANSPARENCIES
Required Equipment and Materials

1 Clear acetate sheets, 8½ by 11 or 10 by 10 inches (21.5 by 28 or 25 by 25 cm)

2 Transparency mounting frames (cardboard masks)

3 Colored pencils which make transparent marks (audiovisual pencils)

4 Acetate-adhering inks and pens for applying them

5 Felt- or nylon-tipped pens (either water-based or quick-drying ink)

6 Pressure-sensitive color sheets, dry-transfer letters and symbols, transparent tapes

7 Overhead projector and screen

Assignment

Plan a transparency for overhead projection that may be used in teaching an important topic in your field. Preferably, design the transparency presentation plan to require at least one overlay. Follow the directions below in making the transparency, and when it is completed, project it for evaluation.

1 *Planning the transparency.* Lay out the complete visual on paper. To determine the correct format area, use a transparency mounting frame. Decide which parts of the visual will be shown in the base transparency and which will be

added with each overlay. It is desirable to make a rough sketch of each overlay on tracing paper; with these roughs you can experiment with the overlays and perfect them before drawing the final transparency.

2 *Producing the transparency.* With the final roughs for your visual, place a clear acetate over the original drawing and trace the base portions of the visual; then, on separate sheets of acetate, trace each overlay. Register marks placed in the corners outside the projection area of each sheet of acetate will aid in assembling the components for the completed transparency.

  Experiment with the following tools and materials to make the base transparency and each of the overlays:

• Water-based felt- or nylon-tipped pens and audiovisual pencils (for very temporary visuals)

• Quick-drying felt- and nylon-tipped pens and acetate-adhering inks (for permanent transparencies)

• Lettering aids and dry-transfer letters and symbols for professional-looking presentations

• Transparent tape for graph lines or borders

• Transparent pressure-sensitive color sheets for large areas of color.

3 *Mounting the transparency.* To a standard transparency mounting frame, attach the base acetate to the *back* of the frame with masking tape; be sure the visual will be face up when the frame is turned over. Hinge the overlay sheets to the front of the transparency frame with masking tape; common practice is to hinge the first overlay on the left edge of the frame, the second to the right edge.

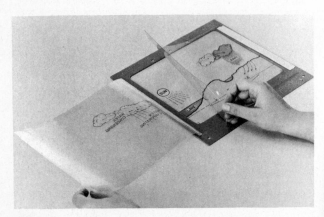

SIMPLE ACTION TRANSPARENCIES

Required Materials

1 Clear acetate sheets, 8½ by 11 inches (21.5 by 28 cms)
2 Transparency mounting frames
3 Assorted small pieces of acetate or plastic
4 Transparency marking pens
5 Split rivets, thumbtacks, pencil eraser

Assignment

Design and make a transparency that has some movable parts or that otherwise uses motion.

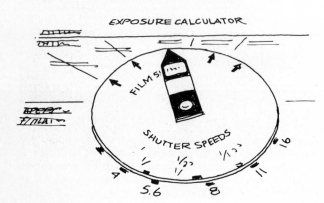

Helpful Hints

• For an axle for a rotating element, push a large thumbtack upward through the transparency base and the moving part; press a pencil eraser over the tack point. Use this technique for spinners, clock hands, figurine arms and legs, and wheels.

• For scenery and backgrounds, add colored plastic pieces and silhouette figures; move for animating a story.

• Diagrams or lines, such as streets, may be placed on the base transparency, and models of automobiles or other symbols may be made from shaped plastic and manipulated from above.

• In graphic displays, colored plastic pieces may be cut to show fractional parts of areas or angles.

• For language study, words, phrases, or clauses may be printed on clear acetate and manipulated on the transparency base.

POLARIZED LIGHT—SIMULATED MOTION

Required Materials

1 Polarmotion Kit or Technamation Kit, polarizing action sheets
2 Manually or electrically operated polarizing spinner for use with an overhead projector
3 Acetate sheets, transparency mounts, acetate marker pens, and art or X-acto knife

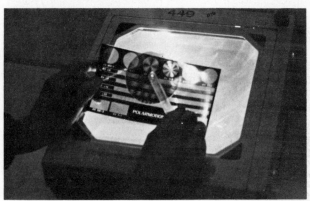

Assignment

Plan, design in detail, and make a transparency with materials to produce simulated motion with polarized plastic.

Producing the Transparency Examine the patterns of polarizing plastic sheets; lay them on the overhead projector table, and look through the rotating spinner to see the patterns. Put your transparency on the projector table, and then select the right pattern of polarizing material to produce the desired motion effect. Trace the desired shape of the motion pattern on the backing of the material, and cut out the form with the art knife. Remove the backing paper and adhere the motion pattern plastic to the base transparency. Burnish with your finger to make a permanent bond. Project the finished transparency with the rotating spinner in the light beam.

EXERCISE 36

PICTURE LIFTS

4 After lifting each picture, and mounting it, project it for evaluation according to generally accepted criteria for transparencies including legibility, satisfactory color, and appropriateness for your purpose.

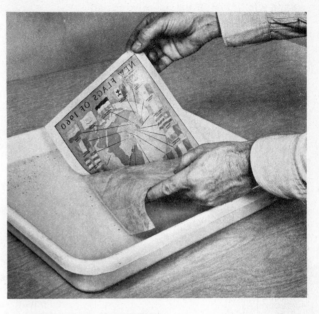

The Reason for It

Making picture-lift transparencies for the overhead projector is an inexpensive way to bring a wide range of pictures into your classroom. The lift transparency requires minimal time and skill for production, and images are bright and in full color. Three techniques for making lifts are given in this exercise.

Before You Start

Brown, Lewis, and Harcleroad: *AV Instruction*, Chap. 7. This manual: Exercise 32; Correlated References, Section 42.

Materials and Equipment

1 Pictures to be lifted. Remember: the lift process removes the printed picture from the page from which it is taken.

2 A pan and water.

3 Clear plastic spray.

4 Clear adhesive film (Con-Tact, or equal).

5 Laminating machine; a dry-mount press; or hand roller.

6 Seal-lamin and Seal Release paper.

Assignments

1 Select a topic which can be presented effectively with pictures on the overhead projector.

2 Collect pictures for making transparencies that present the topic. To be lifted, these pictures must be printed on clay-coated paper such as that used in many magazines of quality. To determine whether the paper is clay-coated, rub a moistened finger on a clear page margin; if a white, chalky residue remains on the finger, the picture can be lifted.

3 Lift your pictures by one or more of the methods presented on the following page, according to the resources available to you.

LAMINATING MACHINE

To make a lift, a roller laminating machine can be used to apply pressure to seal adhesive-backed plastic to a picture.

DRY-MOUNT PRESS

The dry-mount press provides heat and pressure to seal lamination film, such as Seal-lamin, to a picture.

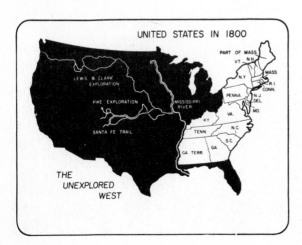

UNITED STATES IN 1800

THE UNEXPLORED WEST

LIFTS WITHOUT MECHANICAL EQUIPMENT

The basic steps in lifting pictures to make transparencies are exemplified in the process by which no mechanical equipment is required. Illustrated to the right, pictures 1 through 6, are the steps for using clear adhesive film such as that used for covering shelves; Con-Tact is the brand shown. Refer to Exercise 12 for a technique that may facilitate application of the adhesive-coated film to the picture: loosen an edge of the backing paper and adhere it to the table top; place the picture in position; then remove the backing as you apply the film to the picture; use a ruler as a squeegee to ensure even application. A hand roller may be helpful to eliminate air bubbles between the picture and film to make firm adhesion.

LIFTS WITH THE DRY-MOUNT PRESS

A Seal product, Seal-lamin or a Laminex laminating film (see also Exercise 12), permits successful picture lifts with any dry-mount press. Use as follows:

1   Place a sheet of ¼-inch masonite board under the press pad. Heat the press to 300 °F.

2   With the picture between sheets of porous paper, predry it in the press for about 45 seconds.

3   Place the picture face up on a piece of kraft paper slightly smaller than the picture. The paper will absorb any residual moisture and prevent bubbles. Cover the picture with Seal-lamin, dull side down.

4   Make a carrier of Seal Release paper—a folded sheet will hold the picture ready to press; the film will not adhere to the release paper. Without Release paper, in step 3 above use a sheet of Seal-lamin slightly smaller than the picture to be lifted so that it will not adhere to the press. Hot-press the picture-plastic sandwich for 60 seconds. Remove and cool it for 60 seconds under a flat weight covering the entire picture. With care to avoid wrinkling it, remove the work from the press.

5   Wash, peel, clean, and finish the lift as shown above.

6   After the picture lift dries, spray with clear plastic spray.

LIFTS WITH LAMINATING MACHINES

The method used to lift pictures with clear adhesive film is the same as that described for lifting without machines; the laminating machine supplies firm, even pressure to the picture-film sandwich. See Exercise 32 and also follow instructions for laminating that accompany the machine. Then soak, wash, fix, and mount the lift.

1   Peel the acetate from the protective backing, or follow the instructions in Exercise 32 for Steps 1 and 2.

2   Place the acetate on the face of the picture, and rub the surface to ensure complete adhesion, free of bubbles.

3   Regardless of the method of sealing—with or without mechanical equipment—soak the acetate-and-picture sandwich in warm water.

4   After 3 to 5 minutes, peel the paper very carefully from the picture.

5   Rub and wash off the residual clay that adheres to the picture on the acetate.

6   After the picture lift dries, spray it with clear plastic spray to make it transparent and to protect it. Then mount for use.

EXERCISE 37

MAKING DIAZO TRANSPARENCIES

The Reason for It

The diazo process is particularly effective for the preparation of crisp, brilliantly colored transparencies from original diagrams or specially prepared masters. Sometimes called the "ammonia process," the method is simple, yet the preparation of overlay transparencies can produce step-by-step presentations of marked value in teaching. Masters can be prepared with minimum skills, using relatively simple techniques, or can be purchased in sets or bound volumes. Commercial masters can often be modified to meet local requirements and then produced as required with locally owned equipment.

Before You Start

Brown, Lewis, and Harcleroad: *AV Instruction*, Chap. 7. This manual: Exercises 20, 21, 22, 25; Correlated References, Section 42.

Purposes

1 To develop skill in planning and laying out copy for diazo transparencies
2 To develop skill in printing diazo transparencies

Required Materials and Equipment

1 Translucent tracing paper, pen with opaque ink, and construction paper.
2 Diazo film of selected color. The film is clear until developed.
3 Ultraviolet-light printer.
4 Developing unit in which to process diazo film. Since

commercial ammonia (ammonium hydroxide) vapor is the developer, it is often used in a covered jar, with a sponge or other device to hold the liquid ammonia.

Although the equipment available for your use might differ from that described here, the basic process of diazo reproduction is always the same.

Assignments

1 Understand the process: For the diazo process, master drawings must be prepared on *translucent* tracing paper, with opaque marks—black ink for words and lines and construction paper for areas to be in solid color. During the exposure of the film, ultraviolet light can pass through the paper, but not through the opaque marks. After exposure, when the film is developed in vapors of commercial ammonia, color appears only in those areas in contact with the opaque marks on the master which blocked the rays of the ultraviolet light. At least five colors of film may be available for your use.

2 Select a suitable instructional topic that can be explained with a transparency. Prepare the master, and then make the transparency as described in the procedure below. Present your finished transparency to the class for evaluation.

   a Prepare a master diagram for tracing paper.
   b Place on the drawing a sheet of diazo film of the selected color. Prepare to expose the two to ultraviolet light in a printer. Note the correct order: light below, master drawing, film on top.
   c Set the timer as recommended, and make an exposure.
   d Transfer the film to the container of ammonia vapor for development.
   e Keep the film in the container until the image completely appears and the color areas are rich and strong.
   f Remove the film, and mount the transparency.

Special Tips

• Be sure that lettering or symbols to be included on a transparency are large enough to be read by everyone in the

MAKING DIAZO TRANSPARENCIES

1 Prepare an original diagram on tracing paper.

2 Place a sheet of diazo film (of selected color) on the drawing. Prepare to expose the two to ultraviolet light in a printer.

3 Set the timer as recommended and make an exposure.

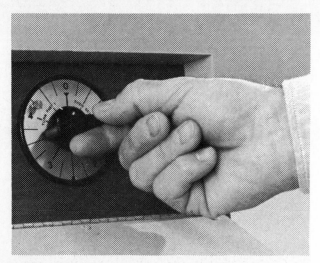

4 Transfer the film to a container of ammonia vapor for development.

5 When the image has appeared fully, remove the film and mount the transparency.

6 With masking tape, mount the transparency on the back of a frame to facilitate projection.

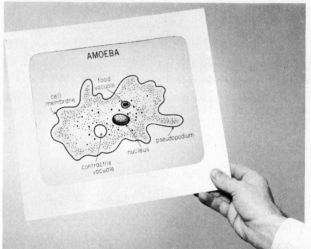

class or audience. A minimum letter size of one-thirtieth the height of the frame area is essential. See Exercises 21 and 22 for suggestions about lettering.

• Keep the content simple. If a large amount of information must be presented, divide the topic into two or more transparencies.

• Divide complex topics into overlay sheets so that different aspects can be presented in sequence. Register each overlay at the corners of the original drawing; then reproduce each on a separate sheet of diazo film. Use different colors as appropriate. Mount the overlay sheets on the frame so that during your presentation each can be flipped into place as needed.

• Plan your transparency so that it fits the frame in a *horizontal* rather than vertical format.

• The quality of the opaque marks on the tracing paper is important in making a satisfactory diazo transparency. Black india ink or special opaque black felt pens are best. Pencils, typewriter, and ordinary felt pens are often not suitable; test before you do all your work.

• If areas of a transparency are to be in solid color, attach construction paper in the proper place on the tracing paper. It will entirely block out the ultraviolet light and result in a rich, even-colored area.

• If the image on the transparency film is *too light*, the film has been *overexposed*. If *too dark*, the film has been *under-*

exposed. Reexpose the master with another sheet of film, and correct the exposure time. For economy, cut a test strip which should be exposed over an important area; expose and rocess; then evaluate results before using full sheets.

While exposure time in the printer is critical, development ime is not. Develop the film long enough to obtain maximum color. Remove the film from the ammonia any time thereafter.

• Practice your use of the overhead projector prior to presenting your transparency before any group. Refer to Exercise 56.

Review Questions

1 Explain in your own words the principle of the diazo process.

2 What material should be used as the working surface for the master?

3 How would you prepare a master that required a large, solid-color area?

4 In what order are the materials placed in the ultraviolet-light printer?

5 Which is critical for time—exposure or development?

6 If the lines on a diazo transparency are very weak, does this indicate over- or underexposure. Would you expose a new sheet of film for a shorter or longer time?

EXERCISE 38

MAKING HEAT-PROCESS TRANSPARENCIES

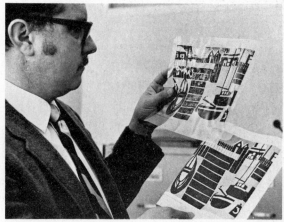

San Diego City Schools, California.

The Reason for It

The heat process is a rapid way to make transparencies. Only a few seconds are required to run a master and a sheet of heat-sensitive film through an infrared heat copy machine. The resulting transparency is completely dry, ready for immediate use. The Thermofax Secretary or similar infrared copy machines are readily available in schools and widely used for making instructional transparencies.

Before You Start

Brown, Lewis, and Harcleroad: *AV Instruction*, Chap. 7. This manual: Exercises 21, 22; Correlated References, Section 42.

Purpose

To develop skills in planning and preparing heat-process transparencies

Required Materials and Equipment

1　Ordinary paper, a No. 1 pencil, and a boldface typewriter with a fresh carbon ribbon.

2　Heat-process film. Black-image film is most commonly used, but color-image films are sometimes desirable.

3　Thermal heat-process copy machine.

Although the equipment available for your use may differ from that described, the principle of thermal reproduction is basic to all heat-process machines.

Assignments

1　Understand the principle of making heat-process transparencies: This process requires that the words and lines on a master drawing be prepared with heat-absorbing material, such as a carbon-base ink or a soft-lead pencil. When the drawing and film pass over the infrared light source, the markings absorb heat, and the resulting increase in temperature affects the film, forming an image on it immediately. The paper surface for the master can be of any type, such as ordinary typing paper.

2　Select a suitable instructional topic capable of being explained with a transparency. In the sequence described on the following page, prepare the master and then the transparency.

3　For evaluation by the class, present the transparency as it would be used in a typical instructional situation.

Special Tips

•　Review the Special Tips 1 to 4 in Exercise 37. They also apply to the preparation of heat-process transparencies.

•　Test-marking pencils are preferred for making opaque marks on the master paper. Black printing inks are also satisfactory, but inks of other colors, most ball-point pens, and purple spirit-duplicated copies are not suitable for thermal reproduction. Clippings from newspapers and other print make excellent masters, but be alert to the small size of lettering that is designed for close-up reading in such materials.

•　If you use india ink or other water-based writing material, make certain the ink is completely dry before the master is run through the copy machine.

Review Questions

1　What is the main thing to check when preparing or selecting a diagram to be made into a heat-process transparency?

2　Is the kind of paper for the master important in this process?

3　If lines on a heat-process transparency are very weak, is the film over- or underexposed? Must you slow down or speed up the machine?

4　Compare the heat process with the diazo film procedure in Exercise 37; compare time, materials and equipment required, complexity of the procedures, and the resulting products.

PROCEDURES FOR MAKING HEAT-PROCESS TRANSPARENCIES

1 Select or prepare a diagram on white paper in india ink, pencil, primary typewriter, or all three.

2 Turn on the copy machine, if necessary. (Some models have an automatic switch that operates when materials are fed into them.) Set control on low or slow setting. (In many machines this will be next to the white oval.)

3 Place the heat-process film, with notch on upper right-hand corner, on the master diagram.

4 Feed the two, together, slowly and evenly into the machine.

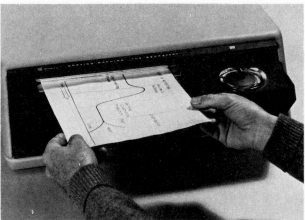

5 Separate master and film as they emerge.

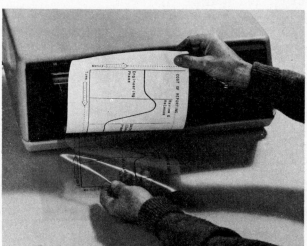

6 Mount the transparency for use.

INVENTING GAMES AND SIMULATIONS

The Reason for It

Games and simulations are assuming ever more important roles in learning. A major reason for this is that teachers and students can produce games and simulations with local needs in mind. Inventing worthwhile games and simulations is a productive learning experience.

Before You Start

Brown, Lewis, and Harcleroad: *AV Instruction*, Chap. 14. This manual: Correlated References, Section 26.

Purposes

1 To learn how to design one's own games and simulations

2 To produce games and simulations tailored to local needs and conditions

3 To test games and simulations by having students use them, and to revise as indicated by results of the tests

Materials

1 Cardboard, white, 22 by 28 inches

2 Standard filing folders

3 Dice and spinners

4 Laminating film

5 Film containers; egg cartons

6 Small beads, paper clips, rubber bands

7 Pregummed colored-paper dots; felt pens

8 Paper cutter, scissors

Assignments

Locally invented games and simulations are most successful if they meet some particular learning need or problem. In approaching the task of inventing games and simulations, or combinations of them, be sure you understand the differences between and requirements of the two types. (See Chap. 14 of Brown, Lewis, and Harcleroad, *AV Instruction*.) Also remember that an important part of the design process is a tryout of your efforts at various stages of development.

The design process itself will almost surely include the following procedures:

• *Write a learning objective.* Develop the basic idea for the game-simulation to help students meet this objective. What drill or practice is to be provided by playing the game? What decisions are to be made by the players that guide them to participate actively? What are the procedures for the game? What do the players do?

• *Set limits for scope and detail in the game.* Keep the objective in mind as you develop one main idea or concept to be the theme of the game. Provide limited drill. Respect time limitations, curriculum priorities, and costs in relation to the benefits of the game for the learners.

• *Identify the players.* What roles will they play? What identities will motivate their participation in the game? How many can play?

• *Determine the criteria for winning the game.* What must players do or accomplish to win? Provide for chance as well as skill so that all players, skilled or not, have some chance of performing with success or winning.

• *List resources required for the game,* such as game board, paper, cards, dice, spinners, questions on paper or cards, markers, and scoring sheets.

• *Determine sequences for interactions.* Who does what and when, and under what conditions? What happens when a correct response is made, or points scored? What happens when an incorrect response is made?

• *Write out the content, instructions, concepts, tasks, questions, and other facts* upon which the game is to be based. Construct the game, or obtain materials for it. Arrange it for use. Play the game and evaluate the outcomes; make necessary changes in the content, procedures, or materials indicated by the initial play.

• *Try to provide a self-check procedure* for students using the game or simulation.

Ideas to Try

You will find most practical "open-end" games that employ a wide range of formats and materials. Change the task cards, the rules, the dice, the spinner; adapt one board for many games. Let yourself go: for materials, adapt egg cartons, film cans, and playing cards. Modify, create, change, invent. Try the following:

• *Use a file folder as a game board.* Use one half for a path, curved and creative. Build in bonus and penalty spaces. Add pictures cut from gift wrapping paper, magazines, newspapers. Use the other half for pockets for task cards or question cards. Laminate the playing surface for long life. Keep rules simple.

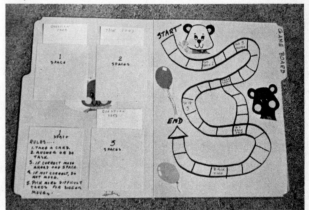

• *Create with egg cartons.* For *Mathematics* number each divider space. Roll two dice. Do something with the numbers. For *Spelling* use the egg carton as a shaker and storage game box. Make cardboard alphabet squares; place a number value on the back. Shake carton. Open, spell words with letters in each section. Develop manual *dexterity.* Design a toss game. Toss chips or corks into numbered or lettered spaces.

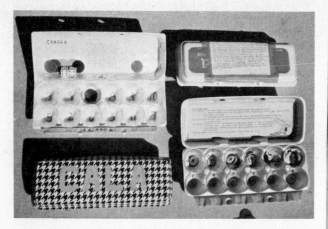

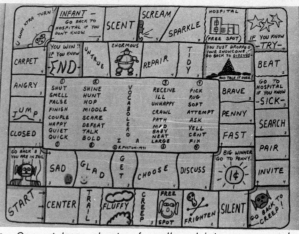

- *Use plastic filmstrip containers for listening skills.* Place various small items in them. Players shake container. Guess at contents. Score equals number correct. Containers also great for small shakers for letters or numbers for toss games.

- *Convert large sheets of cardboard into more complex games.* Use large, colored, pregummed dots (pressure-sensitive, for example) for stops on winding paths. Or adapt the bingo format to meet your game objectives.

- *Cut cardboard into playing-card size.* Design rummy-type games for such subjects as science, nature study, mathematics skills, vowels, verbs, and parts of speech. Laminate for longer life. Or buy blank cards; invent a game.

- *Go Giant Size.* Provide action games on 4- by 4-foot square masonite. Put letters on squares on the board. Have players spell words by tossing bean bags or by hopping from one space to another. Or number the squares. Have players toss bean bags to add up to a selected total. Or, for more activity, hop from space to space.

- *Design and construct a simulation.* Follow same general idea. Use "chance" cards to introduce information, problems, good fortune. Force decision making. Players assume roles, make decisions in line with roles, accept consequences.

WORKSHEET 39

INVENTING GAMES AND SIMULATIONS

Name _____

Course _____ Date _____

YOU MAY COPY THIS FORM TO COMPLETE THE EXERCISE

Name of game or simulation _____

Level (Grade, age) _____ Number of players _____

Item for Evaluation	Players' Reaction (Cite Specifics)	Revisions Recommended (Cite Specifics)
Did players learn the rules quickly?		
Did players remain interested?		
Did interaction process work?		
Did scoring procedure work?		
What learning occurred?		
Other points for evaluation:		

EXERCISE 40

PUPPETS AND PUPPET THEATERS

The Reason for It

Through the creation and use of puppets, information and important skills can be studied and wholesome attitudes developed. Children, as well as adults, are intrigued by the message of the puppet. Communicators in business, the health fields, and education have used puppets to present messages dealing with a variety of subjects such as supervisory techniques, health habits, training objectives, and basic information.

Before You Start

Brown, Lewis, and Harcleroad: *AV Instruction*: Chap. 14. This manual: Correlated References, Section 35.

Purposes

1 To encourage you to use puppetry to provide opportunities for students to express themselves and to participate in an activity that involves much creative construction, as well as presentations to others

2 To provide experience in making simple puppets of several types

Required Materials

1 Styrofoam balls

2 ¼-inch wood dowels

3 Cardboard scraps

4 Felt pens (a variety of colors)

5 Hammer, nails, strips of wood, masonite panels, scissors

6 Paper sacks, crepe paper, and felt

7 Cloth scraps and material for costumes

Assignments

1 After reviewing the references, consider the type of puppet you might enjoy developing into a character of a familiar story or a new, quite individual character. Make some sketches of the puppet as you visualize it; use your sketching technique to try expressions and shapes.

2 Construct the puppet, using techniques pictured on the next page, or use another technique you have located in the references.

3 Plan a brief scene, and present it to some of your associates or to the class. Explain the possible benefits to students from making and using puppets. How could the puppet you made be used with your own students?

Helpful Hints

• Often the feaures painted on puppets are a major factor in their effectiveness. Work out expressions in sketches before applying paint to the puppet. For ears, cardboard may be used, or they can be molded from hard-drying plastic or cut from stiff cloth; even simpler, they may be painted on.

• Newspapers or other paper can be used to make patterns for clothing. Cut out the desired shapes from cloth swatches, sew as required, and fasten to the bottom of the puppet head. Yarn may be used for hair.

• Hands may be made from cardboard or from stiff cloth.

• Sound effects, music, and even prerecorded narration can be developed for puppet presentations.

• Stages can be made from cartons or simple draped frames, or an open closet door, with a draped stick across the opening.

• Two persons or a small group can often evolve an effective puppet-show presentation by planning their puppets around characters involved in a simple plot. Remember, by planning, different types of puppets can be used in the same presentation.

• Have you seen puppets in the theater or on TV lately? The Muppets? Puppets can be for grown-ups as well as for children!

Other Puppet Ideas

Stocking puppets.

Cardboard face fastened to a band on pupil's hand.

Cardboard face on a stick is excellent for lower-grade children.

Silhouettes make good shadow puppets.

Simple Puppet Stages

Plywood, cardboard, or masonite stage with hinges for folding.

A classroom closet with cardboard or cloth stretched across the opening.

making papier-mâché puppet heads

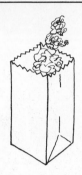

Fill a paper sack with crumpled newspaper and place a cardboard tube in the center of the sack.

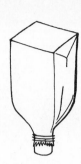

Bind filled paper sack to cardboard tube with string.

Apply strips of newsprint soaked in paste and water. See recipe for making papier-mâché, page 44.

Four to six layers should be adequate.

Paint features, using poster paint. Yarn may be used for hair.

Sew clothing to head. Sew cardboard hands to material.

other puppet ideas

Stocking puppets.

Cardboard face fastened to a band on pupil's hand.

Cardboard face on a stick is excellent for lower-grade children.

Silhouettes make good shadow puppets.

simple puppet stages

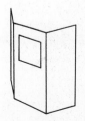

Plywood, cardboard, or masonite stage with hinges for folding.

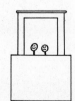

A classroom closet with cardboard or cloth stretched across opening.

EXERCISE 41

MAKING AND EDITING AUDIO TAPES

The Reason for It

Because audio tapes are used extensively, skill in making them and in editing them is important to both teachers and students. Many times, existing tapes need to be shortened or provided with inserted material; sometimes entirely new tapes need to be produced—for example, for music and narration to accompany slide showings or for audiotutorial instruction. The fundamental skills required for making and editing audio tapes are not complex and can be mastered with knowledge and with practice. This exercise is a guide for developing them.

Before You Start

Brown, Lewis, and Harcleroad: *AV Instruction*: Chap. 10; Reference Section 1. This manual: Exercises 52, 53, 54; Correlated References, Sections 3, 4.

Purposes

1 To learn to make an audio tape of satisfactory instructional value and technical quality

2 To be able to edit an audio tape by either the open-reel, two-machine method (electronic) or the tape-splicing method (physical)

3 To learn to make a duplicate of an existing tape on either an open-reel tape recorder, a cassette recorder, or a high-speed tape duplicator

Required Equipment and Materials

1 Two open-reel tape recorders

2 One cassette tape recorder

3 An audio cassette duplicator

4 Supplies and accessories: blank audio tapes and audio cassettes; audio tape splicer and splicing tape; audio tape bulk eraser; audio patch cords, as required for the machines to be used.

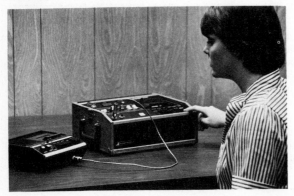

Assignments

1 Using an open-reel recorder and a clean, totally erased tape, make a practice recording of yourself reading three short sentences from this manual. Play back the recording, and evaluate the quality. Rerecord the sentence, if you can improve the recording. Listen especially for noticeable variations in volume, noise of your breathing, or excessive background noise.

2 Remove the second sentence from the recording by making a tape-splice (physical) edit. Replay the tape, and check the accuracy of your edit. If you made an error, repeat the recording and edit it.

3 Rerecord the three sentences; then edit out a word or a few words or a phrase—still retaining the meaning—by using the two-machine (electronic) method.

4 Rerecord the three sentences; then by the two-machine method, edit out the second sentence. In the space of the second sentence, rerecord a new sentence. Check your edit. If you made an error, repeat the assignment.

5 Duplicate the recording made in assignment 4 to a cassette recorder or to another open-reel recorder. Replay a short sequence of the original master tape; then immediately play the same sequence on the duplicated cassette tape; repeat the comparison several times. Evaluate for any apparent differences in the quality of the recording; consider possible causes of any difference in quality.

6 With an open-reel tape recorder, using a microphone, conduct a 5-minute interview with another person. Prepare a 2-minute quotation that can be incorporated in an audio tape for a sound-slide presentation, or a quotation that could be included in a lecture to bring authority to a topical discussion. You may obtain your original material in either of two ways: interview another person who has some special knowledge or background of interest, or record a sequence from a commentator on a radio station. Edit the recording to not more than 2 minutes; include your own *brief* introduction of the speaker and the topic, followed by the quotation. Use what you consider to be the most appropriate editing method.

7 *(Project)* Prepare a short audiotutorial lesson unit. Select a bit of content from a class lesson in a subject of your choice. Design and produce a brief audiotutorial lesson which will include the objective of the lesson and a means of

evaluating learning from the unit. Include in the written materials the list of resources with which the student is to work, any forms or drawings to be followed or worked with—or a description of them, and, finally, your recording to guide the student through the unit.

Live Recording

When recording instructional tapes, and especially when the material to be recorded is to be edited, a number of observations can be helpful:

• For original recording, pronounce words correctly, articulate clearly, and speak in a natural, conversational style. Vocal punctuation is both highly desirable for the listener and helpful to the editor. Such punctuation includes pauses made for emphasis, to permit the listener to absorb the thought, and those made for a change of pace.

• When two or more individuals are being recorded, run some tests to determine proper microphone placement for balancing the voices—loud and energetic, soft and mild. Avoid multiple microphones, if possible. Use the appropriate microphone, i.e., unidirectional for one voice, bidirectional for two, and omnidirectional for several. Try to have performers close to the microphone (6 to 8 inches); with practice, voice delivery can be accomplished without breathing being evident.

• If your recorder has Automatic Gain Control (AGC) or Automatic Level Control (ALC), decide whether this feature is to be used to advantage or whether it will defeat your purposes; often in interviews, the AGC will be helpful, but in dramatic programs where voices and volumes need to be varied for proper effects, recording without the AGC may be best.

Editing Tapes

Editing is an essential skill for preparation of effective instructional recordings. Inevitably, you will want to eliminate portions of a recording, or to combine portions of several tapes, or to make additions to original recordings. In addition to the information you have obtained from Exercises 52, 53, and 54, here are two methods for editing tapes: two-machine and tape-splicing. Each method has advantages for different editing objectives and the type of original material you have available. Be prepared to explain the advantages and disadvantages of each method.

• *Two-machine method.* After planning editing requirements (a marked script will be helpful), set up two tape recorders. Connect the output jack of the machine that will play the original tape to the input jack of the machine that will record the edited version. Start the second machine recording; then start the other machine playing. Stop the second machine to eliminate unwanted portions of the playing tape. (Use the *pause* control, if provided.) Dexterity is required here, and practice is valuable. To avoid thumps and pops in the edited tape, turn down the volume on the recording machine before stopping it; start the machine before turning up the volume.

With this method portions of several tapes can be combined in one recording. Tapes recorded at different speeds can be combined by playing them at their various speeds—1⅞, 3¼, 7½ inches per second (ips)—and rerecording all of them at only one of these speeds. Stereophonic and monophonic recordings can be combined, as well as tapes recorded in different track configurations.

• *Tape-splicing method.* To edit tape by splicing, cut and reassemble the tape to eliminate, add, or combine sections. Segments of tape can be identified by writing on the glossy

side with a felt pen. After a sequence of tape clips has been assembled, duplicate them on a new tape; preserve the spliced tape as a master. Note that when tapes are edited by splicing, each clip must be recorded at the same speed.

Practical Pointers

• Use an open-reel tape recorder for original recording, and record at the fastest speed provided on your machine, usually 7½ ips. The faster the speed, the easier the editing. Have the tape completely erased before recording, and if you use a dual-track machine, record only one track of your tape. Can you explain the reasons for this recommendation?

• When editing, especially when erasing portions of a program and reinserting new material, it is important to use a felt-pen marker on the glossy side of the tape to show the start of the open, erased portion, and the end. To be prepared to activate the pause control, some editors use a dotted line ahead of a cue mark. Remember the erase head is ahead of the record-playback head; marks on the tape should be made at the erase-head position. Tape marking is also helpful in physically editing tape by the tape-splicing method.

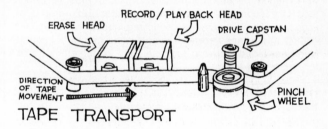

TAPE TRANSPORT

Duplicating Tape

One or two copies of a master (original) tape can be made by the two-machine method, and duplicates can be made from audio cassettes to ¼-inch tape or vice versa, as required. For more rapid duplication, and when multiple copies are needed, high-speed duplicating machines are used; some units duplicate from ¼-inch tape/open-reel masters to cassettes, others from a master cassette to multiple cassettes.

When duplicating audio tapes, remember to make a short test recording in order to set the levels; cassette-to-cassette duplication necessitates adjusting the levels of both tracks appropriately, since both tracks are copied simultaneously. Follow instructions supplied with the equipment.

Do You Know?

When you are employed in instructional activities, you may be interested in developing audio programs for your classes. However, we strongly suggest that you give careful consideration before undertaking new productions and that you investigate prerecorded, commercially available programs that may meet your needs. There are numerous sources of well-produced tapes, and procuring and using them may save many hours of time. To fit prerecorded audio programs into your own instructional design, you may edit them and provide bridging inserts. A *warning*, however: If you buy a prerecorded tape or cassette, be sure to investigate the terms of the copyright and how they bear upon your intentions. You may obtain permission in writing from the copyright owner to edit and use the material as you propose. See Exercise 20, Copyright Law, and *AV Instruction*, Reference Section 2, for further information.

EXERCISE 42

VISUAL LITERACY

Association for Educational Communication and Technology.

The Reason for It

As used in this exercise, the term *visual literacy* implies the ability of individuals to read single pictures and also pictures in sequence as they may read words and sentences; it further implies the ability to analyze pictures and to arrange them in sequences that have meaning for them and that can communicate meaning to others. In the development and use of media, the skills of visual literacy have important application. This exercise is directed toward basic techniques that can improve the skills of young people to observe, organize, communicate, and think.

Before You Start

Brown, Lewis, and Harcleroad: *AV Instruction*, Chaps. 8, 9. This manual: Correlated References, Section 43.

Purposes

1 To explore three elements of visual literacy: *sequencing* (as in stories, processes, and events), *classifying*, and *identifying part-whole relationships*

2 To motivate creative ways of thinking about things, problems, and ideas

3 To explore ways in which pictures influence thinking

Required Materials and Equipment

1 Miscellaneous groups of pictures suitable for the grade level of your students; pictures may be purchased ready-made or may be collected locally:

• Photo-Story Discovery Sets are available at nominal cost from the National Education Association, Publication Sales Section, 1126 Sixteenth Street N.W., Washington, D.C. 20036. There are four sets, each containing from twenty-six to thirty-eight photographs, 3½ inches square, mounted on card stock; two sets are in black and white, two in color. A fifth set —the Visual Categories Discovery Set—is composed of fifty color photographs.

• Pictures from magazines or other sources may be assembled on topics typically of interest to students of the grade level involved; these pictures may be appropriately cropped

Eastman Kodak.

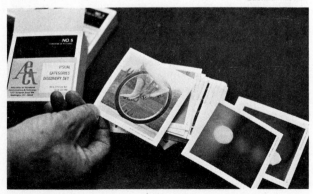

and mounted by techniques explained in Exercises 7 and 8. Criteria for selection of the pictures may be inferred from the assignments below.

2 A still-picture camera, inexpensive, automatic or semi-automatic, that uses Instamatic film. As appropriate, the film used may be for color slides or color prints.

Assignments

Note that, while the assignments below are for young children, materials appropriate for mature students can be selected and prepared for similar experiences. As you evaluate picture sets and try the assignments by using them, first have one of your colleagues role-play a child working with the pictures; later have children do the assignments.

SEQUENCING PICTURES: STORIES

1 Start with a set of a few pictures, each of which shows a portion of a true-to-life event; ask the students to arrange the pictures to represent a story. Each of the sets, 1 and 2, of the Photo-Story series contains several events shown in picture sequences and are suitable for this exercise. Have each of the students arrange a group of pictures and then tell the story to the class. Start young children with a few pictures, and increase the number of picture choices as ability to discriminate between relevant and irrelevant visuals increases.

2 Have a group of students develop a story told by a sequence of pictures. With the picture sequence as a guide, one student may describe the events in the pictures. Students from other groups may ask questions about story details.

3 Using No. 3 of the Photo-Discovery Sets—which has a clear central theme but many pictures that provide numerous story possibilities—have teams work out and display a complete story with pictures alone. Let the groups study the several picture stories and select the story they believe to be the clearest and most logical. Let them also select the clearest and most complete story that is composed of the fewest pictures.

SEQUENCING PICTURES: PROCESSES

1 Show your class—or a visual literacy study group—a set of slides or flat pictues that depicts a process: a step-by-step series of events which results in a satisfactory product or the conclusion of a process. Ask them to point out possibilities of other arrangements of the visuals that might improve clarity of the explanation and to find visuals that could be eliminated without omitting steps necessary to explaining the process.

2 Have each team select a process familiar to them and devise a plan for a picture sequence to teach the process. (Examples of processes are: Sharpening Pencils, Using a Filmstrip Projector, or Finding a Book in the Media Center.) After each of the teams has made photographs or found and arranged photographs and drawings which they believe necessary, have them test the effectiveness of their sequence of visuals by having other students use the pictures to learn the process. Have them evaluate the success of the picture sequence and determine what changes are needed, if any seem to be required.

SEQUENCING PICTURES: EVENTS

1 To give a visual report of an event or to characterize contemporary history, have students select magazine or newspaper pictures that they might put in a time capsule to be opened in 100 years. Have them discuss their reasons for the selections and then arrange them in priority sequence. Finally, have them arrange a display of their final selections with captions.

2 Have students arrive at a statement of events representing cause and effect, such as spring rains–spring flowers. With only a few helps, let them select their cause-and-effect topic. Then let them search for pictures that suggest or depict the entire cause-effect series of events. Have them explain the sequence of occurrences, the time required, a summary generalization of the cause-effect event, and give verification of the accuracy of their conclusion.

CLASSIFYING VISUALS

1 Use the Visual Categories Discovery Set, or similar pictures from other sources; have students group pictures into such categories as plants, animals, colors, squareness, roundness, and things that move. Have them first study the pictures.

2 Have individual students select a category to be illustrated. With a limited number of photographs that each takes —perhaps six—have them display their visualized interpretation of the category. Categories might be found in such words as happiness, work, freedom, frustration, or success.

3 Invite students to write autobiographies with pictures. Limit the number to twelve, or fewer, flat pictures or slides. Permit them to use captions, a recorded narration, or a musical accompaniment.

PART-WHOLE RELATIONSHIPS

1 Have students take a series of pictures of places, buildings, local scenes, or a room—all familiar to the student group. With masks cover each picture as it is shown, exposing only a small spot to give a clue to the area shown. Gradually expose more of the picture until someone volunteers to identify it. Insist that words be used to tell why the scene is recognized. Finally, after several have volunteered identification, expose the picture and verify the judgment.

2 From newspapers, magazines, or other sources, make a collection of pictures that contains pictures hidden within pictures. Have students work in teams to find the hidden pictures most quickly.

3 Have students photograph a series of pictures of an interesting scene or object, starting with a very close picture, then widening the view in successive pictures until volunteers are willing to predict what will be shown in the last picture in the sequence.

PICTURE INFLUENCE ON THINKING

How do pictures influence your thinking? Choose several magazines from which to select pictured advertisements. Try to obtain some from professional, popular, trade, and hobby fields. Then examine each of the ads you selected and consider the following questions about it.

1 Who is expected to read the magazine from which the ads were taken? What can you tell about that person by looking at the ads?

2 What are the subtle, as well as the obvious, messages in the ads?

3 Is the appeal intellectual, emotional, or sensory?

4 How did the composition of the picture attract you to the message? Experiment with the rule of thirds, which says that the center of interest in a picture should not be in the center but off center. Cut four pieces of string and place them over the picture in this pattern: Where is the center of interest placed? How did the advertiser use color, shape, and/or placement to draw your eye to the center of interest?

5 What difference was there in the amount and content of written material in ads from each magazine?

6 Compare ads for the same type of product. How were they similar? How did they differ?

Helpful Hints

• There is extensive literature on visual literacy. See the Correlated References, and seek opportunities to develop and use visual literacy experiences for your students.

• When working with children, start with concrete visual experiences, especially those that help develop concepts of size and space relationships, and order of events (first, second, third . . .) that are necessary before sequencing can be understood.

• Build written and visual communication skills together. With bilingual students especially, explore correlation of visual literacy skills and spoken and written language skills.

• Have mature students plan and produce motion pictures, with either film or video techniques. Encourage them to select topics and scenarios that correlate visual communication, written communication, and subject matter interests. Give opportunity for students to show their creative work for evaluation.

• Resist any impulses to impose your own feelings, judgments, or values upon students as they work to put pictures in sequence, or as they develop slide or motion picture productions. However, urge them to evaluate all their work with honesty and thoroughness.

FILMS WITHOUT CAMERAS

The Reason for It

With very little technical ability and with inexpensive materials, students and teachers may prepare hand-drawn, double-frame, 35mm filmstrips and 16mm motion pictures in the classroom. By making films without cameras, your students learn to understand the fundamentals of a sequential presentation of visuals and the development of a pictorial story. As they verify the accuracy of facts they want to communicate, they can explore the values of library research.

Before You Start

Brown, Lewis, and Harcleroad: *AV Instruction*, Chaps. 8, 11; Reference Section 3. This manual: Exercises 33, 45.

Purposes

1 To develop basic skills in the design and production of simple, hand-drawn 35mm filmstrips and 16mm films

2 To learn that, as your students develop these skills, they are provided with a useful outlet for creative expression

Required Equipment and Materials

1 Drawing paper or printed spirit-duplicated worksheets

2 No. 1 or No. 2 pencils or drawing pens; fine-tip, permanent, felt pen markers of several colors

3 Transparent tape; clear thermal duplication film

4 Thermofax copier; 35mm filmstrip projector; spirit duplicator

5 Rubber gloves and bleach for preparing the 16mm film

6 Used 16mm motion picture film, possibly discarded by an audiovisual center, or a television station; new clear film leader may also be used

35MM FILMSTRIPS

Assignments

1 Examine the illustration, right, and the directions for making hand-drawn 35mm filmstrips. Note that the pattern appears in *full size*. To prepare a master for spirit duplication, trace over lines of the rectangles and lines between rows of frames. Run copies on the spirit duplicating machine. (See Exercise 33.)

2 Prepare copy for 35mm filmstrip on the duplicated rectangular pattern sheets. Go over the frame lines with a soft-lead pencil, or use black ink. After your drawing is complete, duplicate it in a thermal (Thermofax) machine, using heat-process film following directions given in Exercise 38. Complete your assignment by cutting the thermal film into strips, taping the strips together, trimming, and coloring them with permanent felt pen markers. To simplify threading, attach to the head of a filmstrip a 12-inch strip of paper of the same width to serve as a leader. Filmstrip splicing tape or Scotch magic tape may be used.

3 To project the completed filmstrip, insert the leader into a 35mm filmstrip projector as you would an ordinary filmstrip. Since your film has no sprocket holes, project it by pulling the strip through by hand. As frames are projected on the screen, appraise the results.

16MM MOTION PICTURES
Assignments

1 Prepare old 16mm film by the following procedure:

• Soak a 100- to 500-foot film segment in a sink filled with warm water to which has been added a cup of household bleach.

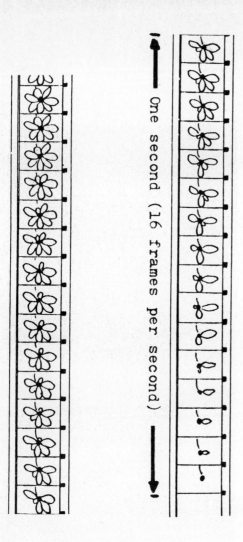

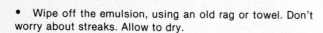

One second (16 frames per second)

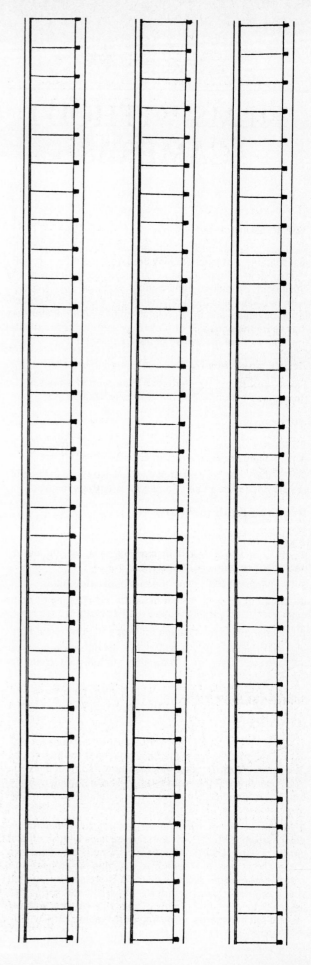

• Wipe off the emulsion, using an old rag or towel. Don't worry about streaks. Allow to dry.

2 Make a spirit-duplicated master from the 16mm pattern at the right, following directions given in Exercise 33. Run a few copies which you will use to develop drawings for your film.

3 With sprocket holes at left, draw or write on the paper masters to develop the sequence you have in mind. Repeat each drawing several times. (See the flower sample.)

4 Cover an area of a table with paper to make a protected work space for transferring your drawings to clear film. Tape a strip of clear 16mm film to the table by using gummed-tape loops fastened to the *underside*. Leave the first 4 feet blank, to serve as a threading leader. Slip the sheet of master drawings under the 16mm film, and trace directly onto the film surface. For best results, use fine-tip, permanent markers or technical drawing pens and permanent inks. Use colors as appropriate.

5 When completed, roll your film on an empty reel, "head out" and with sprocket holes toward the projectionist. Set the projector at silent speed. You may wish to record a musical accompaniment to play during projection.

EXERCISE 44

COPYING WITH A STILL CAMERA

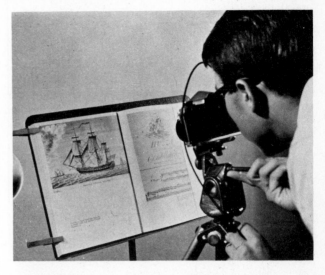

The Reason for It

Widely used in education are photographic copies of pictures, charts, and drawings, as well as close-up photographs of real objects. Equipment required to perform the necessary camera work need be neither complicated nor expensive, As with most other educational media and media devices, imagination and creativity are the essential ingredients for individuals copying or doing close-up photography.

Before You Start

Brown, Lewis, and Harcleroad: *AV Instruction*, Chap. 8. This manual: Exercises 45, 46, 47; Correlated References, Sections 34, 38.

Purposes

1 To become acquainted with equipment, processes, and techniques used in close-up photography and copying

2 To highlight ways in which the results of these activities can be applied to teaching and learning

Required Equipment and Materials

1 A 35mm single-lens reflex camera (such as Pentax) equipped with one of the following: a set of extension tubes, extension bellows, or a macro lens; and a copy stand—preferably one equipped with photoflood lights or electronic flash units. A Kodak Ektographic Visualmaker may also be used.

2 A roll of color-slide film made for artificial light for the camera. If daylight film is used with artificial light, a corrective filter or blue photoflood lights will be required.

3 Sample materials to be copied experimentally, such as charts, calendar pictures, tiny objects, newspaper or magazine print materials.

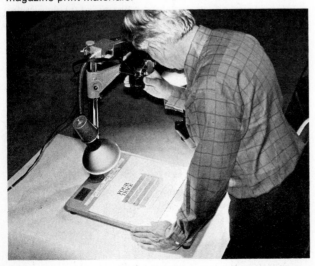

Assignments

1 Prepare a list of objects for which close-up views in 2- by 2-inch slide form would be useful in your special field of interest. Indicate, for each, what the special value of such a slide might be.

2 For the same purposes, prepare a similar list of slides which might be made of maps, charts, textbook illustrations, or other visual materials. Indicate, too, the possible sources of such materials available to you.

Eastman Kodak Company

3 Determine the principal advantages and disadvantages of two widely used copying and close-up photographic systems:

• *Fixed-distance, fixed-area* system employing a Kodak Instamatic camera and copy device which permits copying from set distances.

• *Variable-distance, through-the-lens viewing* which will enable you to copy entire objects or pictures as well as selected parts of them.

Prepare a chart in which to record your evaluations of the two systems. Give attention to the initial cost of each type of equipment; ease of using it; flexibility in terms of size and distance of images, focusing and refocusing, and selecting portions of pictures or objects to be photographed; operating costs for film and light; and size of finished transparencies or negatives.

4 Experiment by copying with one or more of the accessories shown here and compare the results obtained.

• *Supplementary lenses,* placed over the camera lens to enlarge the image and to bring it into focus at close distance.

• *Extension tubes or rings,* used to extend the distance between film and lens, thus permitting close-range, sharp focusing on objects.

• *Bellows extensions,* installed between the camera body and the lens, which permit adjustment of the distance between lens and film with resulting enlargement of the subject.

• *Special macro lenses,* constructed to permit sharp focusing down to within an inch or two of copy.

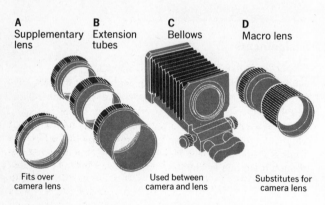

| **A**
Supplementary lens | **B**
Extension tubes | **C**
Bellows | **D**
Macro lens |

Fits over camera lens   Used between camera and lens   Substitutes for camera lens

In doing this assignment, note especially that use of close-up focusing accessories requires increased exposure times or larger lens openings than normal photography. Why is this true?

But, if your camera has through-the-lens automatic exposure control, explain why you may set your camera adjustments for the reading it shows even when using one of the above close-up accessories.

5 Experiment with using a tripod on which to place your camera when doing copying or close-up photography. Place copy on the floor, tack it to a bulletin board, or apply architect's tape to hold it flat against a chalkboard. On the tripod, use a tilt-head device to facilitate composing and focusing on areas of the subject to be copied. What is the effect if the film plane of your camera is not parallel with the material to be copied?

6 Perform a similar set of operations using an adjustable copy stand equipped with two photoflood lights. Experiment to determine results obtained when focusing at various distances between lens front and flat copy; focusing in the same manner on a three-dimensional object; using one or more of the supplementary devices required to make ultra-close-ups; exposing your film with one instead of two photofloods in use; making one exposure using blue photofloods and another with white photofloods. Keep a log of your operations as a means of identifying each of the several pictures made.

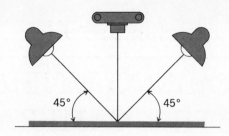

45°   45°

7 Prepare several titles to be copied in slide form, using one or more of the following procedures:

• *Hand-lettering,* following directions for drawing letters and using various types of lettering devices, as explained in Exercise 1.

• *Ready-made lettering,* following directions for selecting and applying letters as presented in Exercise 2.

• *Typewriter lettering,* placed on a 4- by 6-inch sheet of white paper, using capital letters and no more than thirty-two characters per line, double-spaced, and a maximum of six lines of copy per sheet.

Helpful Hints

• Cameras equipped with viewfinders not of the through-the-lens type are not recommended for copy or close-up work because they create the problem called *parallax.*

• Special slide-copying cameras are useful in making same-size or full-size enlargements of selected portions of slides. Some school media centers have this equipment available.

• When photographing titles, be sure to leave ample blank space inside the frame around lettering; select type and copy sufficiently close to produce a screen image that can be read easily by viewers some distance from the screen.

• When copying from books, magazines, or newspaper pages which are printed on both sides, prevent "see-through" effects by inserting a sheet of dark or black paper beneath the page to be reproduced.

• Lighting for copying and for close-ups sometimes presents problems. When accuracy in color reproduction is important, the easiest way to handle lighting may be to do your work outside, in ordinary (preferably midday) sunlight. Photoflood lamps or flash attachments may be used indoors; for color work, make certain that the bulbs are of the type specified on the data sheet packaged with the film you use. For flat copy, adjust lights at a 45-degree angle as shown to provide even light distribution and to avoid hot spots in the photograph.

• When possible, lengthen exposure times and use smaller apertures to increase depth of field, so that the entire image will be in focus.

• Align the camera carefully with the copy so that the lens axis is perpendicular to the plane of the copy. If this is not done, perspective effects will distort the image.

A Word about Copyrights

Individuals who copy materials photographically, as described in this exercise, should be concerned about copyright details. See Exercise 20, "Copyright Law," and Brown, Lewis, and Harcleroad, *AV Instruction,* Reference Section 2, for further information.

EXERCISE 45

MAKING STILL PICTURES

The Reason for It

A camera can make a major contribution to instruction. As you prepare materials for your presentations, you will find many opportunities to use your camera, making records of class activities, pictures of local and distant places and events, selective views of processes and objects to be studied. Such experience will help you master the relatively simple skills of taking pictures and the more difficult processes of determining what pictures to take and how to plan for using them. And as your students learn to communicate with visuals, your competence in photography will enable you to assist them to use the camera for many school purposes and develop skills of visual literacy.

Before You Start

Brown, Lewis, and Harcleroad: *AV Instruction*, Chap. 8; Reference Section 3. This manual: Exercise 44; Correlated References, Sections 34, 38, 40.

Purposes

1   To acquaint you with principal features of several different types of still cameras

2   To help you learn some of the special instructional uses to which these cameras can be put

3   To help you develop elementary skill in choosing subjects and photographing effective still pictures

Required Equipment and Materials

1   Your own still camera, or one borrowed or rented

2   A 2- by 2-inch slide projector

3   Scissors; pen or typewriter (for titles); ruler

4   A roll of film to fit your camera

5   Rubber cement

6   Ten photographs cut from magazines

7   Approximately twenty 2- by 2-inch slides

Assignments

To prepare you to take effective pictures for instruction, here below are several important exercises:

1   Obtain five photographs from magazines or other sources. Assume that it is necessary to reduce the dimensions of each by cropping, at the same time preserving elements that account for teaching values. On a sheet of thin paper laid on the picture, draw lines to form rectangles, squares, or free forms to indicate portions you wish to keep. Try to obtain pleasing compositions as well as to retain essential pictorial content. Adjustable cardboard frames of the type shown in the picture, above, will be helpful aids for this assignment. For each outlined arrangement, write a brief statement that expresses why you believe your cropping has produced a useful picture.

2   Select another set of five magazine pictures that have apparent teaching value. For each write at least two different legends (comments about the picture) and captions (short titles expressing something about content or use). Note instances in which picture meanings or uses may be changed through captions and legends.

3   View approximately twenty 2- by 2-inch slides that demonstrate varying degrees of picture quality from both technical and educational points of view. Rate each on a scale of 1 to 5, with 5 being "superior" or "excellent" and 1 being "of no value" or "poor." In judging the pictures, use the following criteria:

• *Educational usefulness.* The picture is related to some important topic or is valuable in reaching some educational objective

• *Clarity.* The picture message is direct, readily interpreted, and the picture is large enough to be seen when used in the manner for which it is most appropriate

• *Technical quality.* The picture is properly exposed, focused, composed, and appropriately mounted

To express your judgment about each picture, enter your numerical rating and add explanatory comments under each criterion heading.

4 Arrange a meeting of your class or study group with as many as possible bringing their cameras. If necessary, borrow cameras from sources such as the audiovisual laboratory or local camera stores.

Develop a rating chart to be used in assessing at least three different cameras. The features to be compared may include: type of image viewing mechanism; interchangeability of lenses; range of shutter speeds; maximum and minimum lens openings; speed of the lens at maximum opening; type of shutter (focal plane, between-the-lens); automatic features; capability for synchronizing flash; size of film used; capability for producing color transparencies; cost of basic camera with standard lens. Decide whether one of the three cameras will meet your picture-taking requirements; give your reasons. Or, will you need yet a different type of camera for your projects? Give your reasons.

5 Use your roll of film to photograph some operation, place, or event that may be portrayed adequately in the number of pictures the single roll permits. An outdoor location is preferred.

- *Watch exposures.* Pay attention to film speed; light readings; subject motion; position of the sun; the time of day; and other relevant factors.

- *Watch focus.* Focus sharply on the center of interest in the picture. If you are making a close-up, be sure to measure, or visually check in the viewer, to see that the lens opening used will provide sufficient depth of field for the scene to be photographed. Adjust exposure speed and lens opening to produce results desired.

- *Watch your composition.* Work for pleasing compositions as well as for coverage of the picture content you need. If you do not use a camera with a through-the-lens viewing system, be alert to parallax problems. Include enough orientation and close-up shots to convey the visual message. Avoid cluttered or confused backgrounds or foregrounds.

Compare your results with those of others in your group. Assess each set of pictures on the basis of the above pointers. Consider what you have learned about cameras, films, and picture taking through carrying out this exercise.

Helpful Hints

As you shoot your pictures called for in Assignment 5, above, observe the following precautions:

- *Plan your coverage.* Be sure to show essential views, and enough of them, before exposing all of your film. For complex projects, the process of storyboarding—considered in Exercise 25—provides a basis for determining what is to be photographed, the final sequence of the pictures when shown, and the audio accompaniment to be provided: voice, music, and sound effects. In this assignment, however, you are not expected to make a complete storyboard and script but merely to make an outline of your photographic plan.

EXERCISE 46

STORYBOARDING AND DEVELOPING ONE-SCREEN SLIDE PRESENTATIONS

The Reason for It

"The expert knows no more than you do but has it organized and uses slides." This often-quoted remark implies that projected slides can be prime carriers of visual and verbal messages. Images and accompanying sound may present information more clearly and effectively than either medium used alone. By selective emphasis and associated cueing, these two media used together can enrich each other in the communication of precise meanings, impressions, or feelings. Effective slide presentations, however, rarely happen by accident; they are planned, tested, and revised, and the necessary basic skills for preparing them can be learned.

Before You Start

Brown, Lewis, and Harcleroad: *AV Instruction*, Chap. 8; Reference Section 3. This manual: Exercises 41, 42, 44, 45; Correlated References, Sections 29, 34, 37, 38, 39.

Purposes

1 To learn the fundamentals of storyboarding as an aid in planning one-screen slide presentations

2 To use a storyboard and script plan for later conversion to 2- by 2-inch slides as the basis for producing a single-screen sound-slide presentation

Required Equipment and Materials

1 A camera capable of producing processed slides in 2- by 2-inch cardboard mounts

2 A tape recorder, screen, and a 2- by 2-inch slide projector

3 Fifteen or more objects, flat pictures, cartoons, designs, graphs, or diagrams on a central theme or topic. These visuals may be cutouts or original artwork

4 A roll of color film of appropriate length and type

Assignments

1 *Develop a presentation using available flat pictures and graphics.* Select visuals as suggested in Item 3, above, "Re-

All these readily convert to 2 X 2" color-slide medium

quired Equipment and Materials." This assignment may be treated as either a warm-up or a full-scale project. Arrange your visuals into groups and sequences, seeking the best story line to match their visual content and your purposes.

Look for the:

- Best theme, story, argument, concept, or situation
- Logical order for presenting the pictures
- Contributions of each piece to your purpose
- Clear and complete meaning in each visual
- Smooth pictorial development of the idea
- Needed verbal or other sounds required to give the presentation continuity

When pictures are in sequence, number them; also number a file card for each picture. Write your script directly on the cards; say exactly what should be heard as the visual is viewed. Read your script aloud—recording and playing it back on tape is advantageous—and evaluate it; consider such questions as:

- Does the opening establish audience readiness for that which will follow?
- Does what is heard fit appropriately with what is seen?
- Does the narration belabor the obvious elements in the visual, or does it elaborate, clarify, and emphasize the important content?
- Are transitions between visuals smooth, logical?
- Is needed information in the picture and/or script?
- Does the script read easily? Does it sound natural?
- Are there appropriate pauses for analysis of visuals, when required? Are there pauses and perhaps titles for discussion or for students to write or otherwise react to the visuals?

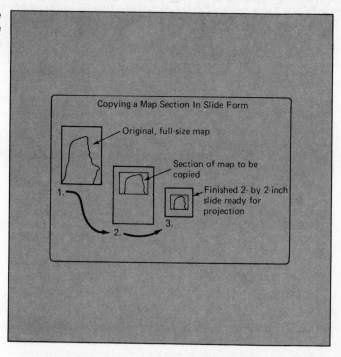

Copying a Map Section In Slide Form

Original, full-size map

Section of map to be copied

Finished 2- by 2-inch slide ready for projection

If you like what you have done so far and have close-up camera equipment, photograph your visuals and proceed as in Assignment 2.

2 *Develop a presentation using slides you already have.* Excellent slide collections are commonly available that can be edited into teaching presentations. If you have such a collection, this assignment is for you.

If possible, state your objectives for the presentation in terms of learner behavior. Determine your approach, your strategy. Will the presentation:

- Overview, review, summarize, teach, test?
- Entertain, tell a story?
- Define a problem, suggest solutions?
- Generate interaction, discussion, activities?
- Advocate a cause or defend a principle?

Arrange and number your slides in the sequence according to your plan. A slide sorter will help.

Provide for sound, using one of the three proposed alternatives for presenting narration and slide change cues.

Provide a script to be read by a narrator, with cues for slide changes marked in the script. Use a script form with headings similar to that in the illustration above. A copy of the script may be required by both the narrator and the projec-

Slide No.	Visual Content	Commentary and Cues	Music and Sound Effects
6	Rommel	// ...continued under the leadership of Rommel and the allied leadership was being urged // by Churchill to open another front...	"Deutschland Uber Alles" up to background (low). . . . fade out as "Victory At Sea" up to low background.
7	Churchill		

Slide change cues

tionist, or slides may be changed by the narrator, using a remote-control unit.

Provide a narration recorded on tape with audible cues in the recording to indicate slide changes. Many sounds can be selected for change cues; try to choose a sound appropriate for the content. Striking a goblet with a pencil eraser, for example, may make a suitable sound.

If equipment is available to permit automatic slide changes by pulses put on the recorded narration tape, experiment with this technique.

Test your presentation before a preview audience; make changes that appear to be needed.

3 (Project) *Develop a storyboard and produce a sound-slide presentation* designed to achieve one or more clearly defined objectives. Start with your objective(s) alone, without prior consideration of visual or script materials. Develop and reproduce in sufficient quantities a data card similar to the one illustrated here on which to record:

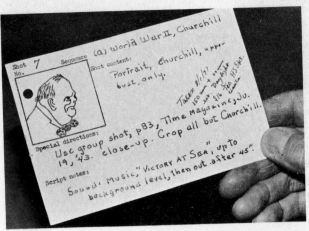

Shot number	Photo data	Narrative (rough)
Sequence number	Visual content	Retake notes
Title	Source	Sound effects
Sequence	Script notes	Artwork

Use the same-size cards of different colors on which to record: the title of the production, credits, short titles or important segments or sequences. Rearrange data cards within each sequence to provide clear direction for the story you wish to visualize. When the arrangement is completed, schedule and take the pictures; edit them for quality, content, angle, and suitability in the sequence. Retake slides, as required.

Using your data cards and your slides as references, revise and rewrite your rough sound script. Test the finished production with an audience, and again revise both slides and sound, as required.

Helpful Hints

• Save time and trouble by using only color film for your slide production.

• Remember that titles can add to the effectiveness of presentations. Titles aid viewers in understanding the organization of your presentation; they may add content. They may also be used to raise questions and to involve the viewer in activities during teaching lessons.

• Often, street signs, newspaper headlines, labels, and other available lettering can be used.

EXERCISE 47

MULTISCREEN, MULTI-MEDIA PRESENTATIONS

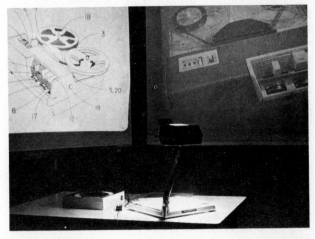

The Reason for It

For presentations that clarify relationships, provide comparisons and contrasts, and create dramatic and highly motivating impressions, multiscreen, multi-media techniques have great potential. Images can be shown in juxtaposition and in sequences; media may be slides, motion picture excerpts, and overhead transparencies. The visuals may be supported by voice, music, and sound effects. For teaching presentations and for communicating with public audiences, multiscreen, multi-media techniques provide a challenge that many students and teachers thoroughly enjoy.

Before You Start

Brown, Lewis, and Harcleroad: *AV Instruction*, Chap. 8; Reference Section 3. This manual: Exercises 20, 23, 24, 25; Correlated References, Section 34.

Purposes

1 To learn techniques of planning and producing multiscreen, multi-media audiovisual productions

2 To develop skill in the simultaneous operation of several different types of projection and audio reproduction equipment

Required Equipment and Materials

1 One or more 2- by 2-inch slide projectors; when two or more machines are used, each should have the same type of slide tray and should be remotely controlled

2 An overhead, 10- by 10-inch transparency projector

3 One or more tape recorder/playback units and blank tape

4 A record player and a selection of records containing background music and sound effects

5 A 16mm motion picture projector; one or more films for scenes to be incorporated as part of the multiscreen, multimedia presentation

6 Two or more projection screens of the same size and surface

Assignments

1 As a warm-up exercise, select several combinations of pictures which can be projected or reproduced by different equipment and which, when projected in that manner, will communicate effectively. Here are several possibilities:

• An overhead transparency of a labeled diagram to be shown along with a slide or a motion picture film clip that shows a piece of equipment, as in the multi-media presentation pictured above.

• A map transparency that can be shown on one screen while related slides (or motion picture sequences) are shown on another; locations of actual scenes can be pointed to on the map as the pictures are presented.

• A transparency showing a topical outline of a lecture presentation while on a second screen still or motion picture illustrations are shown.

• Questions projected one at a time on one screen to direct viewer attention to important elements of a detailed visual appearing on a second screen.

• Simultaneous projection of two or more slides, each showing different views of the same object: oblique overview, front side, back, top, bottom; the display is to clarify relationships.

• Matched photographic slides of chemical symbols and the substances they represent.

• Projection of pages of selections from a play on one screen and stills of actors and actions on another; accompany with dialog and sound effects provided by tape or record.

2 Practice simultaneous projection of several pairs of 2- by 2-inch slides, using two projectors and screens. Note the advantage of using projectors with matching light brilliance

Detroit Library, Michigan.

and lenses of the same focal length and quality. If projectors used together lack equal characteristics, what might you do to compensate for inequalities?

Medium shot   Close-up
Panorama.

To avoid a blinding screen flare after the last slide, put blank slides at the beginning and end of a sequence. Also use a blank slide when one screen is to be dark, or when a picture must appear on it from a separate source, such as a motion picture projector. With this technique, the operator can push both remote-control switches simultaneously for every slide change and also operate other types of projectors.

When projecting slides from the back of the room while simultaneously operating an overhead projector at the front, use the remote-control cord for the slide projector with a cord extension, if necessary. Provide enough slack cord to permit you to move about as well as to operate the overhead projector. Balance light intensity of the two projectors by adjusting the lamp control provided on some slide projectors or by using tinted plastic sheets on the overhead projector.

3 Conduct a test of several ways to control slide changes for a 2- by 2-inch slide projector. Prepare a script for a sound-slide sequence in which these three techniques will be demonstrated:

PHOTOGRAPHIC INNOVATIONS FOR THE CLASSROOM

VISUALS			SOUND
			MUSIC UP

BLANK	BLANK	BLANK	
ALL	GRAND CANYON		
ALL	ROOSEVELT DAM		
ALL	SEDONA CHURCH		
ALL	CAMPUS SCENE		
ALL	PHOTOGRAPHIC INNOVATIONS	FOR THE CLASSROOM	

MUSIC OUT

This presentation covers highlights of photographic processes for learning and teaching, used by the producer over the past thirty years.

ALL	TITLE 2 x 2	BLANK	BLANK
C		FDR	

This section of the program will deal with 2 x 2 slides.

There have been many changes in equipment and photographic materials during this time. However, slides such as these indicate that thirty years ago there were very fine films available.

R		FDR	FLAG
ALL	BLANK	PALO VERDE	BLANK

The color sections of the Sunday newspapers in 1940 were printed in brilliant colors.

Many photographers do not realize that once they have equipped their camera for doing copy work that the same camera can be used for extreme closeups of small objects such as this Palo Verde blossom.

Joel Benedict,
Arizona State University.

• Changing slides on the basis of written cues marked in the narration script.

• Changing slides on audible cues recorded on an accompanying tape; cueing sounds can be made by striking a gong or wood block, tapping a spoon on a glass, clicking a toy "cricket," or by a repeated musical bridge.

• Changing automatically by use of a recorded, inaudible pulse on the accompanying audio tape. Discuss the advantages and disadvantages of each of the three systems.

4 (Project) Select a teaching subject in your field of interest. Develop a storyboard plan for a simple multiscreen, multi-media production on one objective for students in the selected field. Arrange to employ in it a recorded tape, a number of 2- by 2-inch slides, and several large overhead transparencies. Include in your plan sketches of proposed photographs and possible narration, sound effects, and background music. See recommendations in Exercise 46, "Storyboarding and Developing One-Screen Slide Presentations." Use a multiscreen planning form of the type shown below; plan for the use of one or more panoramic scenes, of intermixed titles and graphics, and a simultaneous showing of slides and a related selection of scenes from a motion picture. Now, produce the presentation. Seek artistic unity in the production; make it move smoothly from an interest-capturing opening, through the content to a peak of interest, and to the conclusion. After you have rehearsed your presentation, test it with an audience; invite frank evaluation of its effectiveness and recommendations for ways to improve it.

Many alternatives to a teaching presentation may be selected for this project assignment; typical possibilities include topics for counseling programs, public relations programs, and school or institution activities programs.

MULTI–SCREEN PRODUCTION PLAN

Scene No._____
Time _____

SCENE DESCRIPTION:

Medium _____	Medium _____	Medium _____

VISUAL AND TIMING CONTINUITY:

NARRATION, MUSIC, AND EFFECTS:

Joel Benedict,
Arizona State University.

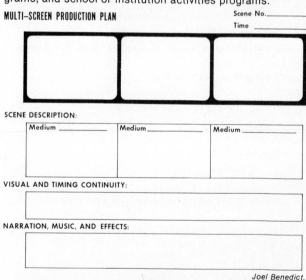

Douglas Aircraft.

Detroit Public Library, Michigan.

102

EXERCISE 48

MAKING MOTION PICTURES: FILMS

The Reason for It

For both group and individual study, school-made motion pictures—especially Super 8mm single-concept types—are often very highly worthwhile. The easy-to-operate Super 8 camera requires only rudimentary skills for filming. Also, with a magnetic projector, adding sound to films is easy. Finally, the film-making process itself provides valuable learning experiences for students.

Before You Start

Brown, Lewis, and Harcleroad: *AV Instruction*, Chap. 8; Reference Section 3. This manual: Exercises 46, 47; Correlated References, Sections 32, 33, 41.

Purposes

1   To identify the features of various motion picture cameras

2   To provide experience in operating a motion picture camera

3   To apply the basic principles of good motion picture photography and editing

4   To provide bases for judging the technical quality of motion pictures

Required Equipment

1   One or more Super 8mm motion picture cameras

2   A tripod with pan-tilt head and floodlights, if required

3   A film viewer-editor with rewinds and a tape splicer for film; 8mm film-splicing tape

4   A Super 8mm projector with or without sound, as required, and a projection screen

Assignments

1   *Examine equipment.* Bring one or more Super 8 motion picture cameras to the laboratory session. Supplement with cameras borrowed from the college audiovisual center or from other sources such as local camera stores. Compare the features and characteristics of the several different cameras. In making comparisons of the Super 8 cameras, complete Worksheet 1 on the last page of this exercise. Practice operating each camera, but become proficient with the one you will be using for your projects.

2   *Study films.* Evaluate sample amateur and professional film productions of an instructional nature. Judge the qualities on the basis of the items listed on Worksheet 2 on the last page of this exercise.

3   *Plan a short film.* Work in groups of three or four persons. Choose a subject suitable for development as a short educational motion picture. Each group should select a different topic. Decide upon the audience for the film, and specify the objectives to be met by using it. Then write a content outline

South Hills Catholic High School, Philadelphia.

of the information that will be treated in the film: describe what is to be taught and what is to be shown. Based upon the content outline, make a storyboard, and finally develop your script. (See Exercise 25.) Apply as many of the motion picture techniques illustrated on the following page as possible. Include titles and, if possible, special effects such as time-lapse, slow-motion, and animation. See Correlated References for specific assistance with these techniques. All groups should compare their plans.

4   *Film the subject.* Load your camera with film and proceed to photograph the scenes indicated in your script. You may find it convenient to film some scenes out of script order. Also, if a scene is incorrectly shot, or the action is bad, re-shoot the scene to get an acceptable "take." Record scene number, take, camera settings (lens setting, distance, camera speed), notes about the action, and concerns to be checked. This "log sheet" will be very useful when editing the film.

Basic Motion Picture Techniques

As you plan to shoot your scenes, keep in mind these basic motion picture techniques:

1   Use a variety of different shots to show a subject.

Long shot (LS)

Medium shot (MS)

Close-up (CU)

2 When moving from one scene to another involving the same subject, carry the movement smoothly. *Match the action* by showing some the same movement in each scene and then removing the overlay action when editing.

3 Do not overuse *moving camera* shots, but do use them when a subject can best be shown with them.

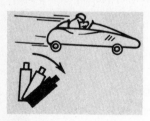

Pan.

Tilt.

Zoom.

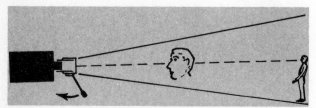

4 Vary the camera angle for some shots.

Low angle

High angle

5 When showing how-to-do-it operations, it is important to place the camera in the position of the learner—the subjective view of the action.

Objective.

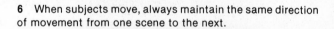

Subjective.

6 When subjects move, always maintain the same direction of movement from one scene to the next.

7 To show the *passage of time* or changes in location, insert brief *cutaway shots* of the subject or other related objects. Cutaways make possible the elimination of long scenes of slow operations or of repeated actions in demonstrations. They are also useful to carry any action smoothly that must be interrupted by stopping the camera. When editing, insert a cutaway scene between two scenes that do not flow smoothly; thus continuity may be improved.

Editing the Film

After the film is returned from processing, project it on a *carefully and thoroughly cleaned projector.* Handle the film with care to avoid scratching it.

Then proceed to edit:

• Remove all bad scenes: poor action, incorrect exposure, or shots that are out of focus.

• Cut scenes apart that are not in proper order, and according to your script plan, splice them together in correct order.

• According to the narration in the script, shorten scenes, and match the action in related scenes.

Most editing is done on a viewer like the one shown below, but upon completion of each step, return to the projector and view the film at the correct projection speed.

Splicing the Film

The most commonly used method for splicing Super 8mm film is with adhesive-backed, Kodak-brand clear plastic splicing tape and a simple splicing unit. Follow this procedure:

1 Align the cut ends of the film on the pins in the splicer.

2 Using the cutter blade, trim the ends so that they match evenly.

3 Place a piece of splicing tape over the pins and the joined ends of the film. Pull off the protective paper on the left side.

4 Pull off the protective paper on the right side.

5 Rub the tape to ensure good adherence. Flip the film over and repeat steps 3 and 4 with another piece of splicing tape.

6 Examine the finished splice for adhesion and registration of holes.

1.

2.

3.

4.

5.

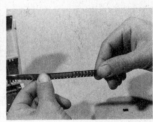
6.

Completing the Film

Check the narration in your script against the edited film. Change the narration as necessary to conform to the scenes. Then, if sound is to be added, send the film to a processing laboratory for the addition of a magnetic sound stripe. When returned, record the sound. Show the film to the class for evaluation.

Name _____

Course _____   Date _____

YOU MAY COPY THIS FORM TO COMPLETE THE EXERCISE

Assignment 1.   Comparison of Features of Super 8mm Cameras

Feature	Camera 1	Camera 2	Camera 3

Lens:
  Fixed focal length
  Zoom type (mm range)
  Focus scale			

Viewer:
  Through-the-lens
  Separate from lens			

Speed selector: Frames per second (8, 12, 18, 24, 32, . . .)			
Footage counter			

Exposure control:
  Automatic setting
  Manual			

Battery-check feature			

Other features:
  Single-frame control
  Adjustable shutter
  Rewind handle			

Assignment 2.   Evaluating Films

View at least one amateur and one professional educational film. Judge the quality of these films on the qualities listed.

FILM TITLES: Amateur _____   Professional _____

Film Quality	Amateur	Professional
Sequential organization of content		
Film exposure		
Camera movement		
Variety of shots used		
Care in matching action		
Techniques used to show passage of time		
Techniques used to show changes in location		
Quality of editing (length of scenes, placement for variety, flow)		
Correlation of narration with pictures		

EXERCISE 49

SINGLE-CAMERA TELEVISION PRODUCTION

The Reason for It

This exercise emphasizes the processes involved in producing edited, single-camera video-taped programs that may be used repeatedly for instruction. In learning the processes, careful attention given to lighting, picture composition, and sound is more than justified. The skills required are easy to learn and easy to teach your students, and the time spent in perfecting them is well invested.

Before You Start

Brown, Lewis, and Harcleroad: *AV Instruction*, Chap. 12; Reference Section 1. This manual: Exercises 64, 65, 66; Correlated References, Sections 21, 33, 41.

Purposes

1 To provide experience in planning simple, single-camera video tape productions

2 To provide experience in producing *edited* single-camera video tapes

3 To present basic principles and techniques of television production, graphics, lighting, sound, and editing

Required Equipment and Materials

1 Single-camera video-tape record/playback system with zoom lens, tripod, and earphones

2 Simple lighting package that consists of not more than three lamps

3 Video recording tape

4 External microphone and cables

5 A two-video-recorder system equipped for capstan servo electronic editing and simple audio mixing

Assignments

While the following assignments could be done alone, we suggest organizing into groups of three or four persons and rotating various tasks such as those of camera operator, video tape recorder (VTR) operator, and performer.

1 Practice lighting: select a person as your subject; set up your camera on a medium shot of your subject; and then one at a time set backlights, key lights, and kickers while observing the effect on a monitor. What effect do different colors of hair have on the amount of backlight needed? Change the height and angle of the various lights, and note the effect on shadows. Use light coming through a window to replace the key lights.

2 Connect the external microphone to your recorder. Have one member of your group read. While monitoring the reading on earphones and recording the test on tape, try various positions of the microphone to determine how far away the performer can be and still have satisfactory voice pickup. Play back the tape, and evaluate results. From which direction does the microphone pick up best? From which direction does it pick up least?

3 Practice editing and cueing by recording a short scene such as a person picking up a cup. Then, using the backspacer scales provided elsewhere, edit a short tape which repeats this same scene six times in a row. Each one in your group should do this editing project.

4 Produce a 3-minute instructional video tape; a topic on some audiovisual technique, such as dry mounting, may be appropriate. Select a member of your group to be the subject matter expert. Write out the instructional objectives and determine the criterion test to evaluate the learning of those seeing the tape.

First, select another member of your group to be the interviewer, and record a "talking-head" interview with your subject matter expert in which he or she fully explains the subject of your tape. Review this interview, and on paper make a list of the specific portions you will use for your narration and the order in which you will use them. Next, list the visual sequences that must be shot to support this narration.

From the interview tape, transfer your selected bits of narration in the correct sequence to an audio tape recorder for later ease of handling. You now have an audio tape which represents the whole narration track of your program.

Now, in sequence, shoot the close-up scenes necessary to support your narration. Be sure to record the natural background sounds of the scene along with the picture.

Edit together your final production. Mix the narration from audio tape with the natural background sound from your recorded close-ups. Play back your final edited tape. Did you achieve each of your objectives? Where and how might you improve next time?

Production Strategy

You may not have all of the tools of Hollywood but you can turn out effective and professional-looking instructional tapes if you keep the process simple. The following outline presents a strategy which allows you to make maximum use of hindsight and to deal with one thing at a time:

1 Define your objectives. Who will be your audience? What do they know now? What problem will be solved with the tape? Is there a simpler way to solve it? How will your audience be different after watching the tape? Be specific!

2 Gather your basic data. Generally there are one or two subject matter experts available who can give a clear verbal description of your topic. Don't ask them to write a script. Simply interview them, and question them until they have told your microphone all you need to hear. Don't worry about the order. Just make sure the description is all there.

3 Review, evaluate, and arrange this interview material until you have the essential comments in the correct instructional order and have discarded the rest. An edited audio tape is a convenient form for this purpose, but notes indicating which segments in which order will also serve. This now constitutes your narration, but it is less stilted and more natural than one based on a written script that is read or acted by nonactors. It is also believable to the audience, and it takes less time overall to produce than a formal script.

4 Based on your narration, shoot supporting visual material. Consider use of graphics and slides as well as live action. *Shoot all scenes long:* have the VTR running at least 20 seconds before any important action begins, and let it run for at least 20 seconds after the action has ended. Also record the natural background sounds along with the picture.

5 Edit together the final program. Your object is to use none of the talking-head pictures—or at least as few as you can. Insofar as possible you want to visualize totally those things being talked about.

6 Evaluate the finished tape. Did it meet your objectives? Show the tape to individuals not involved with the project; were they able to meet the instructional objectives? If it is okay, you can erase and reuse your rough tapes. If it is not, shoot some more, edit the revised version, and retest.

Impose minimum restraints on people appearing before your camera. True, some restrictions are necessary. Small objects must be held stationary in a predictable location. Lighting and camera angle may restrict movement. Sudden, quick moves should be avoided, as they are difficult for a camera to follow. The angle of view of your lenses will also have an effect on the stability of your shots. Camera movements appear smoothest when wide-angle lenses are used. Beyond necessary restrictions like these, encourage people to behave naturally, and train yourself to follow them. Keep the situation calm.

Lighting

Good lighting puts shadows where they will bring out dimensional qualities of the subject and emphasize important elements in the scene. Flat lighting, without shadows, produces dull and often unintelligible pictures. Poor lighting, such as overhead fluorescent light, may cast unflattering shadows

under the chin and eyes. Professionals use a multitude of lighting instruments with many names to produce subtle effects. However, three basic types of light, intelligently used, will give you well-lighted television pictures:

• *Base light* is the overall level of illumination demanded by the camera to produce a picture. In simple settings it is often light that is already available from room lights or sunlight. It provides a base on which lighting effects are built.

• *Key light* is light which produces shadows on the principal subject to bring out its sculptural qualities. Generally this light comes from in front of the subject, slightly to one side of the camera axis, and not more than 35 degrees above it. Sunlight may be used as key by placing the cameraman's back toward windows or toward the sun. Reflectors made of plywood covered with aluminum foil or "space blankets" may be used to redirect sunlight to the proper angle for your subject placement. Arrange reflectors so that people are not required to look directly into them.

• *Backlight* comes from behind the subject and produces an edge of light around the subject which ensures separation from the background.

The basic guideline for all lighting is: *Keep it simple.* Use only enough lights to give a good picture, and remember what is good lighting for one person or situation is not necessarily good for another. A simple change of hair color from blond to black will require an increase in backlight. When a subject fails to stand out from his background, a change of clothing or background from light to dark will often provide the needed contrast. Large areas which cannot be completely lit can be successfully treated by pooling light on key areas. Note light foreground and background with dark intermediate areas in the photo below.

• *Key light* is added to remove eye shadows and provide shadows which will bring out the dimensions of important facial features.

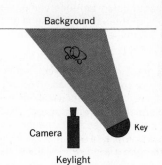

Background

Camera

Key

Keylight

- *Backlight* is added to separate the subject from its background.

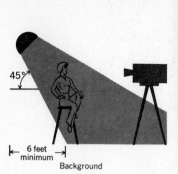

- *Kicker lights:* When the subject is too close to its background, or a single overhead backlight cannot be mounted, two *kickers* on either side behind the subject and out of camera range provide needed separation.

Kicker lights

Audio

Television sound will use the skills you have already developed with audio tape recorders.

As with other aspects of TV production, keep it simple. Use one microphone. Do not attempt multiple microphones until you have totally failed to get an acceptable sound with a single microphone. Multiple microphones multiply problems and require additional equipment and personnel.

Experiment with microphone types and placement. If you have a number of people talking, choose an omnidirectional microphone, and place it centrally in the group. If some have louder voices than others, place the mike closer to those with the soft voices. If there is an unwanted sound at the recording location, choose a mike which has poor pickup from a particular side, and aim this dead spot at the unwanted sound.

Whenever possible, choose a good acoustical environment. Carpet, drapes, and treated ceilings are helpful. Often drapes provided for room darkening can be used to improve room acoustics.

Mechanics of Editing

The minimum requirement for an editing machine is that it be equipped for capstan servo electronic editing. Beyond that, a number of features simplify the process. A "video-only insert" capability makes it easy to lay down the audio and then go back and insert pictures.

During editing, it is necessary to have all the equipment rolling and stable prior to depressing the button. A 10-second roll time is common. This is one of the reasons we have stressed that you shoot all of your scenes *long*. Finding the spot at which you want the edit to occur is relatively easy. Backing up 10 seconds from there can be a problem.

In a video cassette editing system, an editing controller is required and greatly simplifies the process by providing an automatic means of cueing machines. If yours is an open-reel system without an editing controller, a backspacing scale such as that at the bottom of this page will ease the cueing problem. Make a copy of this scale for each machine. Be careful to align the spindle hole exactly; cut out and tape or rubber-cement the scale to the take-up reels of your machines. Copies must be the same size as this original.

To use: Look straight down at the take-up reel and align the outermost layer of tape with the numbers on the scale. You can instantly see how many backward turns of the reel will equal 11 seconds. For example, when the outside layer of tape on the take-up reel is at 6 on the scale, six backward turns of the reel will give an 11-second preroll.

Once each machine has been backed up 11 seconds from the edit point, the two can be started at the same instant and 11 seconds later they will be stable and at the exact point for you to depress the edit button.

Backspacing scale—copy in same size.

Typical editing connection.

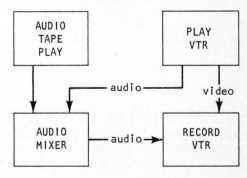

Graphic Materials

Television is a medium which can use other graphic media. Whether or not something can be read or seen clearly depends on a number of variables. Here are a few simple guidelines for graphics to ensure that your message is communicated:

1 Lettering styles should be bold. Condensed letters or those with serifs do not transmit well.

2 To verify that a graphic will be legible, read it from a distance approximately twelve times its width.

3 For normal viewing, twenty-two characters per line is maximum. Long sentences must be broken into several lines for television transmission. Compare the readability of the two frames below from a distance equal to twelve times their width.

University of Colorado

University of Colorado

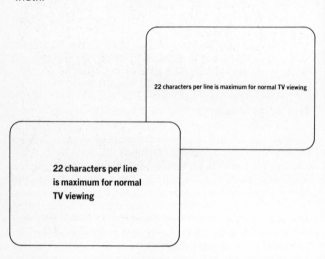

22 characters per line is maximum for normal TV viewing

22 characters per line
is maximum for normal
TV viewing

Advantages of Portable Video Equipment

A principal advantage of television is that it captures events and brings them to students who otherwise wouldn't have an opportunity to experience them. Portability is a must, and the many battery-operated systems are ideal for local production. To increase the mobility of your camera system, try to condense as many parts as possible into a single, compact unit. An inexpensive dolly can be built from ¾-inch plywood and 3-inch casters on which tripod, monitor, and recorder can be placed to permit easy, smooth moves from place to place. Camera movements appear smoothest when wide-angle lenses are used.

Improving Your Skills

Practice is the key to improvement. The more you work with people and the more you work with the equipment, the easier the whole process will become. Critical observation will also help you improve your productions.

Television composition is an extension of developments in graphic arts, still photography, and motion picture films. Look critically at the many examples of these art forms which surround you, particularly television commercials. How does the cameraman arrange one person alone in a shot? How is the arrangement different if the person is looking to the left? Right? Straight ahead? What about two people? When the camera moves, how does the relationship of objects within the frame change? What differences do you find between television and other graphic arts? Why would television rely heavily on close-ups and avoid wide shots?

Observe the results that professionals get from their lighting. View television, particularly the commercials. What kinds of shadows can you detect? Can you detect backlight? Prepare a diagram indicating how many lights you think were used on a specific scene and where you think they were placed.

USING MEDIA IN TESTING

The Reason for It

The inadequacies of purely verbal tests to measure performance and understanding are well known to most teachers. What learners *say* or which multiple-choice alternatives they check in response to verbal problem stimuli may be imperfect indicators of what they know or can do. For the same subject matter, pictorial or three-dimensional tests may have high validity as measures of learning or performance ability because they concentrate attention on realistic, readily understood problems and call for specific responses. In most instructional fields there are many ways to use media in testing and evaluation.

Before You Start

Brown, Lewis, and Harcleroad: *AV Instruction*, Chaps. 1 and 4.

Purposes

1 To emphasize some of the special contributions of audiovisual materials when evaluating results of instruction

2 To provide opportunities to develop and use evaluation instruments that utilize various combinations of audiovisual resources

Required Materials and Equipment

1 Selected still pictures, realia, or equipment on which to base evaluation problems and activities

2 Scissors or paper cutter; rubber cement or cellophane tape; drawing materials and tools; felt pens

3 Stiff paper or cardboard sheets cut to size, approximately 8½ by 11 inches

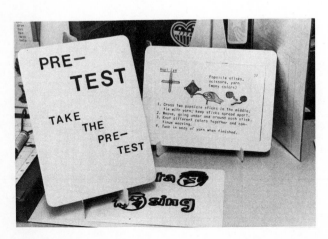

Assignments

Choose one or more curriculum areas, topics, and grade levels on which to base materials developed for the following assignments:

1 Mount on at least ten sheets of stiff paper or cardboard—9 by 12 inches are good dimensions—combinations of pictures, graphs, maps, typing or writing, or realia such as cloth swatches, specimens, or insects—each set of which is to be used as the basis for a multiple-choice question relating to an important learning objective. To guide and systematize student responses, place appropriate verbal comments, directions, numbers, or letters directly on the cards. To determine the clarity and fairness of the intent, wording, and arrangement of each item, try your test(s) with groups of students. Revise items as required.

2 Use your card test items with a group of three volunteers, observing suggestions contained in Helpful Hints at the end of this exercise. Evaluate your success in developing valid test items and in managing the administration of a card test. Try to judge whether the visualized test items aided test takers in understanding the problems and elicited responses expected.

3 Test student understanding of facts or principles by giving a step-by-step demonstration of some process or experiment before the class as a whole. Make correct or incorrect statements or comments as you proceed with the demonstration; students should challenge errors. Demonstration processes can also be carried up to a point and stopped with such questions as: "If I do this, what is most likely to happen?" or "What comes next?" Correct misconceptions and errors in fact.

4 Use motion pictures in similar ways as elements of tests you develop. For example, run a film with the sound off. Then ask students to explain, in their own words, the action that occurred or identities and functions of individuals portrayed. Or run a film to a certain point in the action, stop it, and ask

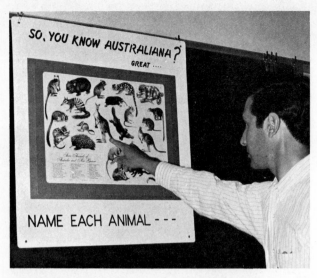

students to predict what is likely to happen next. Or run a film without sound, supply your own explanatory commentary which includes some intentional inaccuracies, and ask students to evaluate your commentary.

5 Experiment with using audio materials as test stimuli. Some possibilities: excerpted *music*—to determine the ability of students to recognize composers, titles, instrumentations, rhythms; excerpted *speeches*—to measure student ability to compare voice characteristics, qualities of arguments, accuracy of facts, delivery techniques; excerpted *dramatizations*—to measure student competence in criticizing performances, to provide a stimulus for students to create their own endings of interrupted actions; excerpted *documentary sounds*—to test student ability to identify animals, birds, coded signals, or other audio effects by sound only.

6 Work in small groups (three or four persons) to develop test problems which employ a combination of realia and chalkboard writing or sketching to measure some important aspect of achievement in a subject of your interest. Assume that each displayed question will be studied and answered by students from their seats and that no verbal directions will be provided other than those written on the chalkboard.

7 Develop a realia test using an audiovisual equipment item, such as the tape recorder unit shown in Exercise 53. Number functioning parts. Frame questions that will measure the test-taker's ability to name parts, to relate parts to functions, and to demonstrate, on a verbal (not operational) level, skills—such as performing steps—quickly and in their proper sequence.

8 Develop a realia test employing at least three biological specimens. Place appropriate directions on cards or on a separate handout sheet to accompany them. Prepare one question about each of the three items and one to be answered through study or comparison of the three specimens as a group. Consider how you might use this same technique in developing tests in your own teaching area. Be specific as to kinds of realia you might use and questions you would ask about them.

Helpful Hints

In administering the types of audiovisual tests described in this exercise:

- Group test items by types of stimuli. For example, all items involving identification of sounds contained on one tape recording should be presented to all students at the same time, in sequence, and each test problem should be identified orally as to number and type of response expected before the stimulus is presented.

- When using card items, prepare a separate question card for each class member. The number of cards with separate questions must be in multiples of the number of students in the class. Thus, a test for a class of 30 will usually require 30, 60, 90, or 120 cards, and so on, unless some students are to be doing other things while parts of the test are underway.

- So that students may answer questions in random order, use standard answer sheets containing enough numbered answer blanks for all items in the test. Then, as the cards arrive in their hands, students may simply match their numbered answers with the numbers on the items.

- To provide a fair basis for timing and to permit use of a standard time interval for all items, try to develop items of approximately equal complexity. In some situations, timed tests are considered appropriate, in which case a time interval is announced for each item, and a warning is given 10 seconds before cards are to be exchanged or the next item displayed, projected, or demonstrated.

- Watch class responses to detect signs of confusion because of speed, or boredom because of too slow a pace. Alter speed accordingly.

25. WHICH ILLUSTRATION SHOWS THE PROPER USE OF A DRILL GAUGE ?

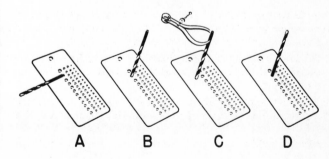

A   B   C   D

EXERCISE 51

SELF-PACED LEARNING: DEVELOPING MODULES

The Reason for It

Since efforts to individualize instruction have led to having students undertake learning experiences independently, and at their own pace, many teachers and media specialists develop packages of instructional materials. For the same purpose, they select commercially produced modular kits that guide students to achieve specified objectives. Criteria used in both selecting and developing kits are similar. This exercise permits exploration of both types of resources.

Before You Start

Brown, Lewis, and Harcleroad: *AV Instruction*, Chaps. 1, 2, 4. This manual: Exercises 12, 19; Correlated References, Sections 27, 31.

Purposes

1 To identify three essential elements of self-paced learning packages

2 To learn to judge the instructional values of several types of modular self-paced learning packages

3 To develop and test your own modular self-paced learning package

4 To be able to identify the specific contributions of each of the various kinds of instructional materials typically included in self-paced learning packages

Required Equipment and Materials

1 Several examples of different types of self-paced learning packages prepared by teachers, such as Learning Activity Packages (LAPs) or Self-Learning Modules (SLMs)

2 Samples of commercially published self-paced learning packages and mini-courses, including at least one of an audiotutorial (AT) type

3 Audiovisual equipment required to audition or view the materials selected for or provided in kits or packages

Assignments

1 Examine samples of several, at least two, different study packages; these may be either commercially produced or locally prepared. Preferably, include one of each type. Appraise each package, using questions in the worksheet for this exercise.

2 (Project) Select a topic in your subject area and grade level. Include the three essential elements described below, and select appropriate commercial instructional materials and/or produce necessary items not otherwise available. Field-test your package, preferably with a sample of typical students for whom it is designed; or have two or more members of your class serve as your test sample. Evaluate the results and, guided by them, revise the program; then retest and evaluate.

In developing this independent-study project, after you have selected a topic or skills that are to be learned or developed, follow the suggestions presented in Developing a Self-Paced Learning Package. Keep in mind characteristics of the student group for whom you are designing the package. Refer to textbooks and audiovisual materials for information you may want to include. Organize the content in a sequence you believe to be appropriate for the students, or design an approach that gives them options in their procedures.

DEVELOPING A SELF-PACED LEARNING PACKAGE

Before you undertake to develop the self-paced learning package described in Assignment 2, read the following descriptions of desirable characteristics of the three principal parts: (1) behavioral objectives, (2) activities, and (3) evaluation.

Behavioral Objectives

Remember that behavioral objectives are statements to students of what they are to learn or what they will be able to do after successfully completing appropriate activities required by the module or package. Your topic or subject may require satisfactory achievement of one or several objectives. If you find you are writing a large number of objectives, consider dividing your topic into several parts, and treat each as a separate package.

Each objective you write should contain three elements:

- An *action* verb: to *make*, to *arrange*, to *identify*, to *mark*

- The *content* reference: the *subject matter* to be treated

- The *performance standard:* the minimum acceptable *level of accomplishment;* if none is stated, 100-percent accuracy or completion of work is assumed

For examples of objectives and sources of information on writing them, see the Sample Self-Paced Learning Package, following, and Correlated References in this manual.

Activities

Now select activities for students that will help them to accomplish the objectives successfully. Activities can include reading of texts, periodicals, pamphlets, and other sources, listening to tape recordings; viewing audiovisual materials—graphics, still pictures, filmstrips, slides, or Super 8mm film; handling real objects; performing lab work; or having other actual experiences.

Develop explicit directions for the students to follow in performing the activities. This is the Study Guide. Provide for frequent practice or opportunities for the students to apply what they have learned by testing their understanding or new ability. Include worksheets and self-check questions. Be sure students are continually informed about how successful they have been.

Evaluation

Develop test questions, direct experiences, or other measurement techniques which relate specifically to each objective. Advise the student to obtain the test or perform the activity upon completion of work for all objectives in the study package.

Finally, test each package, and analyze results; if student accomplishment of any objective does not reach your expectations, consider the instruction unsatisfactory, and revise that part of the package. Revise and retest until your students achieve an acceptable level of successful performance.

SAMPLE SELF-PACED LEARNING PACKAGE (PARTIAL)

Topic: Overhead Projection
Student Group: Teacher-Education Students

SAMPLE BEHAVIORAL OBJECTIVES

1 To plan and prepare at least two kinds of transparencies, each of which receives an average rating by other students of at least 3 on a 5-point scale.

2 To operate the overhead projector well enough to receive a rating of "Satisfactory" by the instructor.

SAMPLE ACTIVITIES (STUDY GUIDE)

Objective A

1 Read text, "Teaching with Transparencies," and "Creating Transparencies."

2 Complete Part 1 of Worksheet. Check answers in back of Manual.

3 Select one process to make a handmade transparency according to procedures described in the Manual.

4 If desired, view the 8mm single-concept film illustrating the technique you have selected. The film is shelved above the projector at the side table and has the same title as indicated in the Manual.

5 Prepare the transparency.

6 Practice using it with the projector in the lab.

SAMPLE EVALUATION

Objective A:
Rating Scale for Evaluating Transparencies

	1	2	3	4	5
Treatment of content					
Use of transparency techniques					
Technical quality of transparency					

Objective B:
Criteria for Using Overhead Projector

_____ Setup and placement of projector in room

_____ Prefocus

_____ Position of user

_____ Adjusting image (elevation, focus)

_____ Handling transparencies

Sample Worksheet

1 What handmade transparency process have you selected?

2 List the materials you have used in preparation.

3 What reproduction equipment process did you select?

4 List the materials and equipment required.

WORKSHEET 51

SELF-PACED LEARNING: DEVELOPING MODULES

Name _____

Course _____ Date _____

YOU MAY COPY THIS FORM TO COMPLETE THE EXERCISE

Evaluate self-paced learning packages according to the following criteria. (Write your answers on a separate sheet of paper.)

1 Are objectives stated in behavioral and measurable terms?

2 Do the items of evaluation directly measure the outcomes of the objectives?

3 Does the Study Guide include sufficient instructions for the student to work through the package without the teacher being present?

4 Is the content divided into short segments, each of which treats one objective or a single concept?

5 Is there only one path for all students, or are multiple activities included to serve students on different intellectual and experiential levels?

6 Are the activities interesting, varied, and appropriate to the objectives?

7 In addition to printed materials, have appropriate media been included?

8 Is there provision for frequent practice, application of knowledge, or other student self-evaluation of content learned?

9 Following each self-check, is the student immediately informed of the results?

10 Have you tried out the package with potential students? What were the results?

SECTION THREE

OPERATING AUDIOVISUAL EQUIPMENT

Many instructional and communication situations require demonstrations of skills related to coordinated uses of audiovisual *materials* and *machines*. The manner in which one carries out such coordination will either impede or help communication and instruction.

Almost everyone has had the experience of seeing slides shown that are out of focus, that spill off the projection screen to the adjacent wall, that are upside down or backward. Or probably you have seen films projected when the accompanying sound was unintelligible. You may have heard disc or tape recordings played in a manner that made either understanding or enjoyment impossible. And television pictures are frequently seen in classrooms with pictures that twist or roll with garbled sound; at worst, picture and sound are missing completely because no one can find the "off-on" switch.

Often, such faulty performance is conveniently blamed upon equipment failure. But in many cases the real cause of the trouble is operator failure: a plug is not pushed in, a dial is not turned, a proper sequence of actions has not been followed in setting up the equipment for operation. The operator has "operatoritis"—fortunately, a curable disease.

Exercises in this section are intended to accelerate, guide, and simplify your learning to operate essential items of audiovisual equipment and to help you avoid or be cured of "operatoritis."

Reminders on Learning Skills

It is a good idea to first reconsider several basic reminders with respect to learning to operate audiovisual equipment:

- *Practice is essential.* Practice, to be of optimal value, should be distributed in a regular pattern. Short sessions spaced throughout a week are usually better than long sessions conducted irregularly.

- *Try to understand the results you wish to achieve.* As you gain general understanding of the requirements of your skills, you will begin to discover and understand new details of required procedures; you will benefit by observing persons more expert than yourself; you will identify cause-and-effect relations; and you will benefit from asking questions and from reading.

- *When you practice—THINK.* Look for similarities in the requirements of two or more skills. Understand *transfer of training or learning.* Use what you have learned in one skill to help learn another with similar requirements. Seek to learn *principles* of operation that also apply to new situations.

- *Appraise your progress; be aware of your growing competence.* Performance checksheets are provided at the end of this manual for numerous types of audiovisual equipment. They permit you to check your performance in such critical areas as effective equipment operation, best sequence of operational steps, completeness of steps, speed of operation, and, when appropriate, some corollary operations. When you feel you are ready, use these checksheets to appraise your progress. Your instructor may set up a test schedule.

- *Return at intervals to each piece of equipment.* Test and reinforce your learning.

Evaluating Yourself

Here is a set of criteria to help you evaluate your own evolving skills. They are adapted from those developed by Dr. Edgar Dale, professor emeritus, Ohio State University. He says skill in operating audiovisual equipment can be divided into four broad categories:

- Unconscious inefficiency
- Conscious inefficiency
- Conscious efficiency
- Unconscious efficiency

A description of each of these categories may help you measure your performance and progress.

- *Unconscious inefficiency.* You are not competent to any degree, either in speed or quality of performance. Worse, you don't know what you don't know.

- *Conscious inefficiency.* This is a sign of growth. You are aware of your incompetence, and you are beginning to understand the points at which your inefficiency is greatest; thus, you know what to study and practice, and why.

- *Conscious efficiency.* This is a very desirable level of achievement and probably the one reached by a majority of serious students during a brief course. You have mastered the principles and technical skills of equipment operation, and you make few errors. But you must make a conscious effort to take all necessary actions and to make all required checks during your performance. Probably your conscious effort will limit your speed, and although you are accurate, your performance will not be as rapid as desirable. Need for continued practice is indicated.

• *Unconscious efficiency.* You are an expert. You make every necessary adjustment and check, you work at high speed with assurance, and you work almost automatically and without errors. You are able to think of other things while you carry on your equipment operation, you can talk with others, plan your next comments, and so on. This level is achieved by much practice, by complete understanding of equipment and operating requirements, and—again—by practice.

Some Further Pointers

Have you noticed the similarity between filmstrip projectors and motion picture film projectors? They are similar in focusing and framing; the film travels in much the same way through both machines, even though one may be turned by hand and the other by a motor; and the audio portion of a motion picture machine has controls like those on a record player: off/on/volume/tone. Other suggestions that may help you gain operating skill rapidly and effectively are:

• *Start with the familiar.* You can turn on and adjust a radio or a television, and you probably operate some record players with considerable assurance. These machines include controls basic to many audiovisual devices: school record players, public address systems, tape recorders, and the sound section of motion picture film projectors. Projection machines, no matter what type, make, or model, operate with some specific principles common to all.

• *Build on your previous knowledge.* Continue to seek relationships between the machines you have learned to operate and new machines you must learn.

• *Obtain competent guidance.* It is much harder to unlearn wrong practices than to learn correct procedures. Attend carefully to demonstrations, hints from your instructor, and films and other resources designed to help you learn the right way to operate machines.

• *Maintain a sequence of steps in operating machines.* There are important reasons for using a well-defined sequence of actions in learning to operate or in operating audiovisual equipment. You may be inclined to debate details of sequences suggested in this manual, the standards of performance required of you, or the possibilities of better, more efficient methods or sequences of operation. You may notice that an expert may use some sequence other than that suggested. But, remember, an expert often combines his actions, or changes their order for good reason; if you are not

an expert, you will improve your skill and speed your learning by following a sequence of checkpoints that (1) help you include every important action, (2) improve your technique, (3) protect materials used in the machine, and (4) require you to double-check your work.

• *Work for both skill and precision.* If you require ten minutes to prepare a motion picture film projector for operation, you have not developed adequate skill; you will waste class time and embarrass yourself. Remember: your objective is to work for *both* speed and precision—not for speed alone, not for precision alone.

Helping Your Own Students

Your ultimate goal in mastering operation of audiovisual equipment may be to prepare yourself to train your own future students. Remember the precepts suggested here; provide situations and procedures by which your students can learn the same skills as those you have developed.

Section Three Exercises

Exercises in this section of the manual direct experiences that lead to proficiency in operating (and, in many cases, to providing first-echelon care for) the following types of audiovisual equipment: audio disc players, open-reel audio tape players, audio cassette recorders, overhead projectors, 2- by 2-inch slide projectors, sound-slide viewers and projectors, filmstrip projectors and viewers, opaque projectors, 16mm motion picture projectors, 8mm motion picture projectors, television receivers, video players and recorders, portable video systems, and microcomputers.

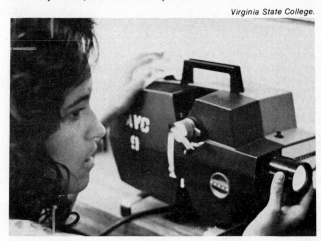

Virginia State College.

EXERCISE 52

AUDIO DISC PLAYERS

The Reason for It

In many courses and for many activities at all levels of education, phonograph records have continual value. Moreover, many audio disc players for instructional use have various features that make them useful as public address systems, as devices for simulating radio broadcasting, and as units in audio systems for making instructional programs. The object of this exercise is to help you obtain the best possible results from audio disc players and to extend the life of your records.

Before You Start

Brown, Lewis, and Harcleroad: *AV Instruction*, Chap. 10; Reference Section 1. This manual: Exercises 5, 41, 53, 54; Correlated References, Section 16.

Purposes

1 To guide you in learning to operate audio disc players
2 To indicate how to care properly for audio disc players
3 To provide practice in the operating skills necessary for the activities in which audio disc players can be used

Required Equipment and Materials

1 School type of audio disc player, especially one with microphone input and control
2 Disc recordings of music, sound effects, and speech
3 Microphone

Assignments

1 Set up and prepare the audio disc player for use.

2 Locate and identify operating controls and determine the function of each, using the form, on the next page, as your guide.

3 Play a recording of music, and test the effects of each control on the player.

4 Plug in the microphone; use the player as a public address system; determine the effect of volume and tone controls; determine the effect of the location of the microphone in relation to the speaker(s); check to see whether the tone control can be used to reduce feedback noise.

5 Put the machine away in proper condition for the next user.

Operating the Audio Disc Player

In completing the assignments above, observe the following good practices. Refer to the performance checklist on page 167. Have someone test your performance before asking to be tested by the laboratory supervisor or instructor.

1 Set the machine on a level table. Open the case, and uncoil the power and speaker cords. First, insert the speaker plugs in the proper receptacles; with a stereo machine, plug in *both* speakers. Always plug in the speakers before plugging in the power cord and turning the unit on. Turn on the amplifier.

2 Select the proper turntable speed for the record selected.

3 Select the proper stylus: LP for 45- and 33⅓-rpm records; standard for 78-rpm records. This is a most important adjustment to protect your records and your stylus.

4 If the weight of the pickup arm is adjustable, set the weight to match the stylus you are using. See the operating guide for your machine. Records are played with the least pressure on the stylus that will track reliably.

5 Turn on the turntable motor, lift the pickup arm from its rest, and place the stylus gently in the run-on groove at the edge of the record. If your machine has a pickup arm lift and cue device, always use this means of placing the stylus on the record.

6 Adjust volume and tone controls, and balance control for stereo, if provided. For the records you have selected, experiment to find the best combination of settings for voice and music.

7 To avoid scratching the record when the record is through playing, allow the stylus to travel into the spiral, run-out groove before lifting the pickup arm *straight up*. Lift the arm with the lift-cue device, if provided.

Special Tips

1 Always handle disc recordings by the edges; never touch the grooves with your fingers.

2 Keep records in dust covers except when playing them. Clean records mean extended life and maintained quality of sound.

3 Never be rough with audio disc players either in use or in transit.

4 Do not play stereophonic records on a monaural player unless the equipment instructions say it is safe to do so. However, monaural LP records may be played safely on a stereophonic machine.

5 Always use the correct stylus for the record to be played.

6 Some machines have a built-in stroboscope which provides visual evidence of correct speed adjustment. Learn to use this valuable aid.

Putting Away

1 Turn off all switches.

2 Set speed control lever at a position to disengage the drive roller from the turntable.

3 Lock the pickup arm on its stand to prevent damage to stylus and cartridge.

4 Store all cords on brackets or in storage compartments in the case. Never coil cords under the turntable; the drive mechanism could be damaged.

5 Secure lid latches carefully and completely.

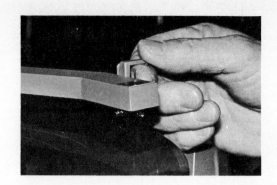

Using an Audio Disc Player as a Public Address System

To use the audio disc player as a public address system, set it up as you would to play records. Plug the microphone into the jack on the control panel labeled Microphone. Adjust the microphone volume by using the Microphone Volume control. Adjust the tone control(s) for optimum clarity and natural quality. To avoid feedback, place the microphone some distance from, and behind, the loudspeaker(s).

On a machine such as that shown below, separate controls are provided for record play volume and for microphone volume. Practice adjusting these controls to provide background music for speech or to insert speech between recorded excerpts.

WORKSHEET 52

8MM MOTION PICTURE PROJECTOR CHECKLIST

Name _____

Course _____ **Date** _____

---

assignment 2. identification form

On the record player you are using, locate the components and controls listed below. As you find each one, identify it by writing a number by each term in the list, and write the appropriate number on or near the item on the drawing below (see example: "Turntable"). You may need to add items to the generalized drawing as well as terms to the list. If items in the list are not on your machine, mark these terms with a 0.

1. Turntable
   Pick-up arm
   Pick-up arm rest
   Stylus selector
   Volume control(s)
   Tone control(s)
   On-off switch
   Microphone jack(s)
   Speaker jack(s)
   Speaker
   Power cord
   Speed selector

EXERCISE 53

OPEN-REEL AUDIO TAPE RECORDERS

The Reason for It

While compact cassette audio recorders have replaced open-reel, ¼-inch tape recorders for many purposes in instruction and training, open-reel machines are still preferred for some purposes. For audio production work, for example, open-reel recorders are preferred because of the ease of editing and splicing ¼-inch tape and because open-reel recorders are convenient to use in combination with other audio devices.

Before You Start

Brown, Lewis, and Harcleroad: *AV Instruction*, Chap. 10; Reference Section 1. This manual: Exercises 5, 41; Correlated References, Section 17.

Purposes

1 To learn how to make and play tape recordings on open-reel audio tape machines

2 To learn to repair and edit ¼-inch audio tapes

3 To learn to use open-reel tape recorders in combination with other audio instruments—record players, radios, and cassette tape recorders

Required Equipment and Materials

1 Open-reel, ¼-inch tape recorder with microphone, take-up reel, and power cord (several models, if available)

2 Recording tape, leader/timing tape (optional)

3 Splicing tape, tape-splicing block, single-edge razor blade or scissors, felt-tip pen that will mark the back of the tape

4 Other audio machines and connecting cords

5 Manufacturers' manuals for each machine used

Assignments

1 Become familiar with controls and other components found on most tape recorders. Study the generalized recorder on the next page. On a machine available to you, locate items identified on the drawing.

2 Following the suggestions in this exercise, set up the recorder, make a short recording of your voice, and rewind and play back the recording. Determine the following:

3M Company.

• Can you use the microphone to start and stop the machine?

• Can you locate precise spots in your recording for replay? Can you use controls such as rewind, fast forward, microphone off-on switch, pause, and the index counter?

• What provisions are made for connecting other equipment?

• Are any features peculiar to your machine alone, such as the means for changing tape speeds? Look for special instruction on the machine.

3 Connect one open-reel recorder to another, using jumper cords. Then edit a voice recording, using the pause control on one machine to omit unwanted portions in the final tape being recorded on the second machine.

4 Make a recording with another person. Experiment with different placements of the microphone; vary the microphone distance from an inch or so to several feet from the speaker. Note the sequence in which you record at the various distances and, on playback, determine the effects.

5 With pieces of scrap tape, practice tape splicing. Compare the free-hand method, using scissors (see next page) with the splicing block method, using a single-edge razor blade. Make a recording of three sentences; edit out the second sentence by splicing the tape; play the tape.

6 Practice using a tape recorder with other audio equipment, including another tape recorder, a radio, and a record player.

A Generalized Tape Recorder

Though all tape recorders have essential features in common, the location and exact appearance of these features will differ among machines. To find these features on the machine you are using, examine the top, back, or sides. Also notice differences in the control keys, push buttons, levers, or knobs for the machine functions. On each open-reel model look for instructions that indicate special features.

Sound and Power Controls

• *Sound controls* for adjusting tone normally function on playback only; volume controls may function on both recording and playback, while some machines also provide automatic volume control (AVC) or automatic level (ALC) when recording.

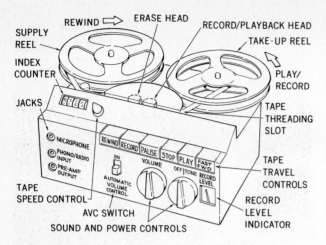

SUPPLY REEL
REWIND
ERASE HEAD
RECORD/PLAYBACK HEAD
TAKE-UP REEL
INDEX COUNTER
PLAY/RECORD
JACKS
TAPE THREADING SLOT
MICROPHONE
PHONO/RADIO INPUT
PRE-AMP OUTPUT
REWIND RECORD PAUSE STOP PLAY FAST FWD
TAPE SPEED CONTROL
ON AUTOMATIC VOLUME CONTROL
VOLUME
OFF TONE RECORD LEVEL
TAPE TRAVEL CONTROLS
RECORD LEVEL INDICATOR
AVC SWITCH
SOUND AND POWER CONTROLS

- A *power* switch may be incorporated in the volume or tone control.

- A *record level* indicator, when provided, may also double as a battery-level indicator for machines that operate on batteries.

- An *automatic volume control* switch may be provided to turn off the AVC feature to permit manual adjustment of recording levels with the *volume* control.

Tape Travel Controls and Indicators

- Controls for tape travel on all machines are *forward* (or *play*), *rewind*, *record*, and *stop*.

- A *fast-forward* control permits skipping program material at a faster-than-play speed.

- A *pause* control, when pressed, stops tape travel, and when released, allows the tape to continue as before.

- An *index counter* aids in locating program segments on the tape.

- A *tape speed control* selects the speed of tape travel for playing and recording.

Input and Output Jacks

A *microphone* jack is provided for microphone input. Two (or a compound jack) are provided on stereo recorders. A high-level input is also provided, usually marked Radio/Phono, to receive signals from an amplified source, such as another tape recorder, a radio, or a phonograph.

Two outputs are common. One is called Preamp or Accessory Amp, which is low-level, used to connect the tape recorder to another recorder or amplifier. The other, a high-level output, is used to connect the tape recorder with a separate speaker or earphone. This output is usually marked Speaker.

Operating Procedures

The procedures that follow apply to most open-reel tape recorders. If possible, refer to instructions for the specific machine you are using:

1 Open and set up your machine. Keep vents clear.

2 Plug in the power cord, and turn the machine on. Set the tape speed control (3¾ or 7½ ips).

3 Place the empty take-up reel on the spindle which turns when the *play* or fast-forward controls are used.

4 With the recorder controls in *stop* position, thread the tape into the threading slot. Attach the end of the tape to the empty take-up reel.

5 To play back a recorded tape, operate the *play* control. If your recorder has a *pause* control, note how it works during playback and recording. Adjust volume and tone.

6 To make a recording, connect the microphone to the input provided. Use both the *record* control and the *play* control to start the recorder. Speak into the microphone. Adjust volume.

7 When you have completed recording, use the *stop* control. With the tape taut between the reels, operate rewind control.

8 Play back the recording. Adjust volume and tone.

Splicing Tape

To edit tape by cutting and assembling sections, or to repair tape that has been accidentally torn, splice the ends as follows:

1 Lap two ends of tape, then cut at a 45-degree angle by holding scissors diagonally in relation to the overlapped tape.

2 Butt the ends together smoothly, then place a short length of splicing tape (such as Scotch Brand No. 41) over the joint.

 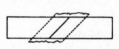

3 Trim the edges of the splicing tape so that no sticky surface extends out from the edges of the recording tape.

 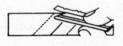

UNINTERRUPTED RECORDING TIMES

The length of *uninterrupted* time which can be recorded on an open-reel tape recorder depends upon

Tape speed: The most common tape speeds are 1⅞, 3¾, and 7½ inches per second.

Tape length: The length of tape which a reel can hold depends on the tape thickness, as well as the reel size.

Reel size, in	Tape length, ft	Tape speed, inches/second		
		1⅞ ips	3¾ ips	7½ ips
3	150	15 min	7½ min	3¼ min
4	300	30 min	15 min	7½ min
5	600	1 hr	30 min	15 min
5	900*	90 min	45 min	22½ min
7	1,200	2 hrs	1 hr	30 min
7	1,800*	3 hrs	90 min	45 min

* Thinner than standard tape.

AUDIO CASSETTE TAPE RECORDERS

3M Company.

The Reason for It

Compact cassettes—a universal standard for convenient, portable audio recording—are used everywhere for training and education, for personal and institutional communication, for high quality music, and for recordings for audiovisual presentations.

Before You Start

Brown, Lewis, and Harcleroad: *AV Instruction*, Chap. 10. This manual: Exercise 53; Correlated References, Section 17.

Purposes

1 To learn how to make and play tape recordings on typical cassette tape recorders

2 To learn how to use microphones effectively with cassette machines

3 To learn some uses of cassette recorders in combination with other audio equipment

4 To learn to use cassette recorders in carrels and listening stations

5 To use the cassette recorder to prepare an instructional tape

Required Equipment and Materials

1 Cassette tape recorder and microphone

2 Cassette tapes

3 Earphones, multiple outlet box for headphones, remote speaker, interconnecting cords

Assignments

1 Examine one or more cassette machines. Attempt to locate the items in the list below, and answer the questions about them. Most of the items will be found on all cassette recorders.

GENERALIZED CASSETTE RECORDER

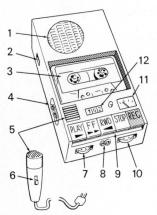

1 SPEAKER
2 EXTERNAL POWER
3 CASSETTE
4 INPUT/OUTPUT JACKS
5 MICROPHONE
6 REMOTE OFF/ON SWITCH
7 VOLUME CONTROL
8 MICROPHONE JACK
9 TAPE TRAVEL CONTROLS
10 TONE CONTROL
11 LEVEL INDICATOR
12 INDEX COUNTER

• *Power supply:* Does the machine use AC power or batteries? Both? Is there an AC power cord or a jack for an AC to DC adapter?

• *Cassette compartment:* How is it opened?

• *Tape travel controls:* Look for forward (play), fast-forward, stop, rewind, record, and pause controls.

• *Microphone and monitor (earphone) jacks:* Are there any other input or output jacks?

• *Volume control, tone control:* Does one of these incorporate the power switch, or is amplifier power turned on by the *play* control, or by a separate switch?

• *Record level meter:* Does this indicate battery level also?

• *Microphone:* Is there a built-in microphone? Does the extension microphone have an *off-on* switch for remote control?

• *Tape position indicator:* When a recording is made, set the indicator at 000, and keep a log of the number at the start of each recording. Information on the tape will be easy to locate by reference to the log.

2 Play a prerecorded tape; then make and play back several test recordings, following the procedures in this exercise.

3 Plan and produce a short teaching tape to be used as part of an instructional unit. Write an objective and plan a teaching sequence and evaluation; then develop a recording for the lesson. Test the lesson with students and revise the tape as needed.

Playing a Cassette Recording

1 If using AC power, plug in the power cord. If using batteries, check their level on the meter by turning the recorder on; press the *play* control if there is no power switch. If the meter does not read within the recommended range, replace the batteries or use an AC adapter.

2 Press the *eject* button to open the cassette chamber. Insert the cassette so that the exposed tape will contact the recorder heads and the full reel is on the left. Remember that the cassette plays from left to right, in the same direction that you read its label.

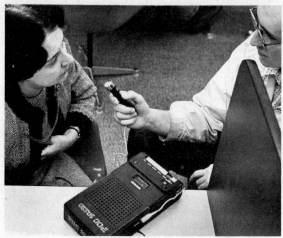

Portland Community College, Oregon.

3 Press the *play* control, and adjust volume and tone as needed.

4 Operate the *stop, fast-forward,* and *rewind* controls.

5 Listen to the tape played through the recorder speaker, through an earphone connected to the earphone jack, and also through an extension speaker (see Exercise 53). Connect the recorder to a multiple headphone outlet box to see if it will drive several headsets.

6 Rewind and then remove the cassette by pressing the *eject* button.

Making a Recording

1 Insert the cassette. Any material previously recorded on the tape will be erased as the new recording is made.

2 Attach the microphone. Note whether it has an *off-on* switch for remote control.

3 To begin the recording, you must use both the *play* control and the *record* control. If there is a switch on the microphone, this must be *on* also. If the recorder has a lock-in *pause* control, you can simplify starting by pressing the *pause* button, then setting the *record, play,* and *microphone* controls for recording, and pressing the *pause* control again when ready to start the machine.

4 Adjust the volume control so the needle swings within the recommended range on the meter. Make a short test recording to check the setting. Many cassette machines have automatic volume control (AVC) for recording. With these, sounds of different levels are automatically recorded at nearly the

same volume. If your recorder has AVC, experiment with recording voices at different distances. Also observe how background sounds are picked up in the recording.

Make a recording using a second tape recorder or other audio machine connected to the first with a jumper cord. Experiment with volume and tone controls on the playing machine to obtain optimum clarity. Special resistance cords may be required for using some cassette recorders with other audio equipment.

Cassettes

Small size, ease of use, and worldwide uniformity in playing speed and packaging have encouraged use of cassettes. The cassette is, in fact, a reel-to-reel device. The tape ends are attached to small hubs enclosed within the case which can be turned over to reverse the direction of play without handling the tape.

Most audio cassettes are played at 1⅞ ips. Cassettes are commonly available in uninterrupted playing times of 15, 30, 45, and 60 minutes. They are designated by the total playing time of *both* sides (C-30, C-60, C-90, and C-120). A C-30 tape, for example, plays 15 minutes *on each side*. Cassettes in special short lengths can be ordered and are often useful in programs for instruction. The C-120 (2-hour) tapes are extremely thin and are recommended *only* when uninterrupted playing time of 60 minutes is imperative.

Examine the two tabs in the back edge of the cassette. These can be removed to prevent accidental erasure of program material. There is one tab for each side of the cassette; remove that tab that is on the left as you read the cassette label for the side you wish to protect (side A, on the cassette pictured). To record on cassettes from which tabs have been removed, cover the holes with thin plastic tape.

Always use good quality cassettes and tapes to avoid broken or jammed tapes and poor recordings. While cassette tapes can be spliced, considerable program material may be lost when a tape breaks.

Monaural and stereo tapes can be played on any cassette machine. Only a stereo machine will reproduce sound separation from a stereo tape.

Using Cassette Machines

What features make these machines especially suitable for individualized instruction and independent study? In what ways would you use them in instructional packages combining audio materials with projected visuals? Worksheets and textbook assignments? Laboratory experiments? Consider these applications as you complete Assignment 3.

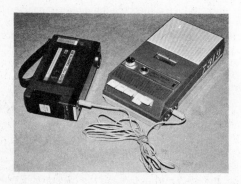

EXERCISE 55

PROJECTION PRINCIPLES

The Reason for It

Projection of instructional slides, filmstrips, motion picture films, and opaque materials involves many simple, fundamental principles that apply to the use of all projection devices. Ultimately, mastery of these principles, and practice in handling individual pieces of equipment, will make you a skillful projectionist.

Before You Start

Brown, Lewis, and Harcleroad: *AV Instruction*, Reference Section 1.

Required Materials and Equipment

1   Glass-beaded screen

2   Matte-surface screen

3   One 2- by 2-inch slide

4   2- by 2-inch slide projector

5   Two objective lenses of different focal lengths to fit the slide projector selected. Most slide projectors come equipped with a 5-inch focal-length lens; try to get, in addition, a 3-, 4-, or 7-inch focal-length lens

6   A slide projector with a zoom lens

Assignment

Read through this exercise and perform the experiments, using the 2- by 2-inch slide projector (see Exercise 57). Other exercises in the manual will help you learn how to apply the same principles of projection when using other pieces of equipment.

Essentials for Effective Projection

To be sure of getting a clearly visible image every time, a projectionist must do four things:

1   Operate the projection equipment correctly.

2   Project a sufficiently large and bright screen image.

3   Place the screen and the projector where all viewers can see to the best advantage.

4   Keep equipment clean and in good condition. These needs can be met by following systematic procedures whenever setting up and using projectors and screens. The projection principles described in this exercise apply to all projection equipment. Refer to later exercises for instructions on operating and maintaining specific types of projectors.

Effects of Projection Distance and Lenses on Image Size

Good projection includes filling the screen. This means that the projected picture is large enough for the viewers to see yet fits on the screen without extending beyond the edges. The following experiments illustrate how a correct-size image is obtained:

1   Place a projection screen at the front center of the room. Set up the 2- by 2-inch slide projector about half the length of the room away from the screen. Project a slide. Does it fill the screen? Is the image of satisfactory size?

2   Move the projector closer to the screen and again project the same slide. What happens to the size of the projected image?

3   Move the projector to the back of the room and again project the same slide. What happens to the size of the projected image?

4   Leaving the projector in the back of the room, replace the projection lens with one of either shorter or of longer focal length. Now project the same slide. What happens to the size of the projected image?

5   Use a projector with a zoom lens. Adjust the lens. What happens to the picture size? Is there an advantage gained by using a zoom lens? Explain why.

Permanent wall screens simplify the use of projected materials. When making a selection, consider type of screen surface, screen size, and placement. The average conventional classroom will require a screen approximately 72 inches wide. A rule of thumb suggests that viewers be seated no closer to the screen than two times the width of the projected image, and no more than six times the width of the image from the screen.

Suggested Screen Sizes

See complete tables of screen sizes for various types of projectors located elsewhere. Note that square screens are often recommended for 2- by 2-inch slide projectors and overhead projectors. Many instruction rooms are provided with one wall painted a matte surface white upon which one or more images can be projected for multi-media presentations.

Measure or arbitrarily specify the size of two instructional areas. Using the tables inside the back cover of this manual, provide the recommended projector-screen distances and screen dimensions for satisfactory viewing when using the following projectors:

1   2- by 2-inch slide projector with 5-inch lens

2   16mm sound film projector with 2-inch lens

3   35mm sound filmstrip projector with 5-inch lens

4   10- by 10-inch overhead projector with 22-inch lens.

Decisions about screen placement and size for the projection to be done are clearly evident in this use of overhead projection in a typical classroom.

Charles Beseler Company.

Lining Up the Screen and Projector

For the entire area of an image to be sharp, the axis of projection must be perpendicular to the screen. If not, the image will keystone and will look like one of these:

Projector too far to left.   **Projector too low.**

When one side of the projected image is taller than the other, move the projector to the right or left until the image is correct. If the image is wider at the top than at the bottom, tilt the screen forward at the top, or pull it toward the wall at the bottom to correct the distortion. Tripod screens often have an arm extending toward the room from the top to make the correction. Position the screen sufficiently high for all viewers to have a clear, unobstructed view.

Proper Brightness

To a large degree, the type of screen surface used determines the brightness of the projected image and the proper seating arrangement required for each viewer to see a satisfactory picture on the screen. Do the following:

1 Experiment by setting up matte and beaded-glass screens side by side. With the slide projector on and focused, but without a slide in it, project the light partially on one screen and partially on the other. Stand immediately behind the projector, and determine which screen gives the brighter image.

2 Now, move as far to one side of the room as you can. Which screen is the brighter?

3 Move along the side of the room, checking every two or three steps toward the front to determine which screen gives the brighter image.

4 What generalizations can you make now about the relative value of the matte and the beaded-glass screens? Under what conditions would each of them be best to use? Or is there any difference? If a lenticular screen is available, repeat Steps 1 through 3, using it for comparison.

5 Move the projector closer to the screen. Does this affect the brightness?

Proper Seating

The above experiments illustrate the difference in screen brightness between matte and beaded screens. Good seating area is between two and six times the width of the image projected on the screen. A beaded screen has a slightly narrower satisfactory seating area than a matte screen.

Screen Placement in a Room

Placement of the screen affects not only the optimum satisfactory seating area for viewers, but also the optimum brightness of the image. If a room is completely darkened, a satisfactory image on the screen is relatively easy to obtain, but if light falls on the screen from any source other than the projector, a degraded image will result.

Refer to the room diagram below in answering the following questions. Assume that the seats are stationary but that viewers can turn.

1 With complete window darkening and a solid door, where should the screen be placed?

2 With satisfactory darkening, which screen—beaded or matte—would provide the larger desirable viewing area?

3 If the room cannot be well darkened, possibly because window drapes leak light, where should the screen be placed?

4 Draw pencil lines on the diagram to show the desirable seating area when using a matte-surface screen.

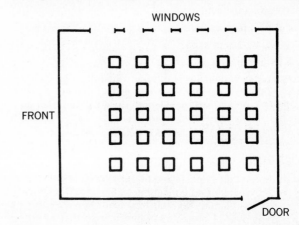

Screen Placement: 20- by 30-ft Room—Matte Surface Screen

Consider placing a screen in a corner of the room for improved image brightness and seating convenience.

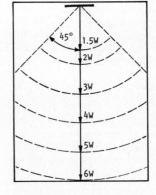 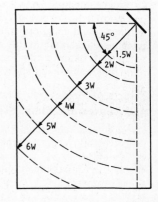

124

EXERCISE 56

OVERHEAD PROJECTORS

The Reason for It

For group instruction, the overhead projector gives complete and direct control of an enlarged image on a large screen in a lighted room. At the same time, the instructor may face the class, watch reactions, and respond to questions. Transparency materials are easily prepared and may be used over and over again. Overhead projection is widely used for communication in business, industry, government, and education at every level.

Before You Start

Brown, Lewis, and Harcleroad: *AV Instruction*, Chap. 7; Reference Section 1. This manual: Correlated References, Section 19.

Purposes

1 To learn to use large-transparency overhead projectors
2 To become acquainted with some of the advantages of overhead projection

Required Equipment and Materials

1 Overhead projector
2 Projector stand
3 Projection screen
4 Demonstration transparencies, including polarized samples

Bell & Howell.

5 Audiovisual marking pencils, china marking pencils, and felt pens

Assignment

Although several models of large-transparency projectors are on the market, all operate in the same way. The machine pictured below is normally used in presentations to very large groups; the smaller machine shown is typical of many classroom models. Practice setting up and using the overhead projector, following the procedures described below.

1 Set the projector on a suitable stand (overhead projector stands are usually lower than other projection stands). Plug in the power cord. Wrap the cord around the stand leg. Why?

2 Turn the projector on. The cooling fan will operate whenever the projector is on. In many machines the fan is switched on and off automatically by a temperature control; when the projector switch is turned *off*, the fan will continue to run until the unit is cool. Never disconnect the power cord until the fan has stopped running. Why is this practice important?

3 Place a transparency mask on the projector table. Move the projector forward or backward to fill the screen with light. Focus until the edge of the lighted area is sharp. Does the projected image have a keystone form—wider at the top than at the bottom? Placing the projector on a high stand will reduce the keystone effect, but the projector then may be in the line of vision of some of the audience. Try tilting the top of the screen toward the projector. Some screens are designed to permit this adjustment. Raise the front of the projector with a strip of wood from ⅜ to ¾ inch (1 to 2 cm) thick. Try various amounts of keystone distortion. Consider how much distortion you would permit for various projection conditions and transparency content.

4 Use the mirror tilt knob, or tilt the head of the machine to raise or lower the beam of light on the screen.

5 Project large transparencies, using the demonstration materials. Make all necessary adjustments to ensure a sharp picture of appropriate size. Practice pointing to, covering, and uncovering portions of the transparencies, and manipulating overlays.

After completing the assignment above, experiment with the following techniques:

- Write or print directly on clear acetate sheets or, on some machines, write on the plastic material provided on a roller attachment. Determine optimum letter size for readability. Use felt pens, china marking pencil, and audiovisual colored pencils. What are the advantages and disadvantages of each?

- Try progressive disclosure techniques. Place a sheet of opaque paper *under* the transparency or plastic roll material —so that you can see the image—and reveal portions of the image sequentially by removing the paper in stages. For what instructional purposes might this technique be especially useful?

- Project as silhouettes small flat objects, such as hand tools, scissors, templates, gears, pulleys. Can you suggest uses for this technique?

- Cut pieces of assorted shapes from clear and colored plastic sheets. Arrange them to form pleasing designs. Does this technique suggest participation exercises in which learners identify shapes, colors, designs, amounts, sizes, relationships?

- Use the overhead projector to reduce a large chart or drawing to a size suitable for a transparency, as described below. Set up and turn on the projector. Focus and adjust the light beam to cover the exact portion of the large chart that you want to reproduce in reduced size. Set up photofloods or spotlights to illuminate the chart. Turn off room lights and the projector lamp, but keep the spots on. The reduced picture will appear on the projector stage where transparencies are placed. Trace the picture on white translucent paper. Later, prepare the transparency (see Exercises 37 and 38).

- Project motion transparencies. Attach overhead motion adapter, or use a manual spinner, with polarized (Polarmotion or Technamation) transparencies. What are the advantages of this type of transparency?

(American Polarizers, Inc.)

Maintenance

Keep lenses and glass surfaces clean. Use lens tissue only. With the projector disconnected from power, learn how to replace the projection lamp. See further instructions in Exercise 68. Many commercially produced transparencies are available for various subjects at different educational levels. Here are two examples published by Vemco in Big Springs, Texas.

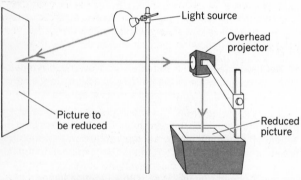

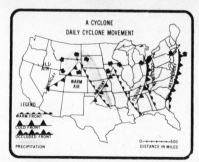

550AA04 basic, overlays 1, 2

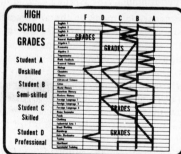

371BO08 basic, overlays 1, 2, 3, 4

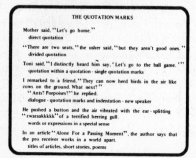

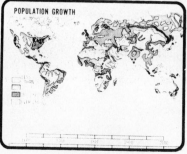

900AHX13 basic, overlay 1

126

EXERCISE 57

2- BY 2-INCH SLIDE PROJECTORS

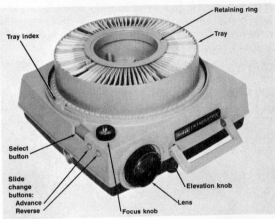

Eastman Kodak Company.

The Reason for It

For either individual or group viewing, 2- by 2-inch slides can be projected in effective sequences. For groups, slides shown on a tray-loading projector can be paced by a lecturer by remote controls, or slides may be changed and a narration and sound effects presented by a magnetic tape. Slide projectors and viewers can be useful also for presenting visual portions of individual study units.

Before You Start

Brown, Lewis, and Harcleroad: *AV Instruction*, Chap. 9; Reference Section 1. This manual: Exercises 55 and 58; Correlated References, Section 20.

Purposes

1 To learn to load slides in slide carriers and trays so that the projected image will appear correctly on the screen

2 To learn to operate single-slide and tray-loading projectors

3 To learn to use special slide projector features and accessories

Required Equipment and Materials

1 2- by 2-inch slide projectors, including some that are tray-loading and that can be operated with remote-control cords or synchronized signals from a tape recorder

2 A projection screen

3 Slide holders, including rotary trays, straight trays, and a stack loader, as available

4 A series of slides

5 Slide projector accessories, including a zoom lens, sound synchronizing equipment, and filmstrip adapter, as available

Assignments

1 Select and arrange for showing a series of a dozen or more slides. Load and project these in several slide projectors, including both manual single-slide projectors and those using trays. Study the instructions for loading below; note how slides must be positioned to ensure that the picture will be projected correctly on the screen.

2 Practice and master the use of special features and accessories, such as remote-control cords for changing slides and focusing the projector, a zoom lens, and automatic slide advance by timer or by inaudible signals from a record player or tape recorder. Make notes of difficulties that arise; determine how these problems in projector operation can be avoided.

3 Prepare a tape to demonstrate the available types of slide programming recorders, and determine the advantages and disadvantages of the slide-change pulse methods used.

Loading Slides

To load slides into slide carriers or trays, position the slides as follows: (1) Face the screen. (2) Hold the slide so that it reads normally. In most cases, commercially processed slides will carry the company trademark on the side of the slide that faces the projection screen. (3) Keeping the same side of the slide toward the screen, invert the slide so that the image is upside down, and insert the slide into the slide carrier or tray slot.

Thumb spots help orient slides for projection. Traditionally, the spot is placed in the top right corner on the side that is away from the screen when the slide is in position for loading.

Before loading trays of eighty-slide capacity for the Carousel projector pictured above, make certain the elongated slot on the bottom of the tray is engaged in the notched starting position (aligned with Zero on the ring) as shown. After loading a Carousel tray, seat and lock the retaining ring to prevent slides from falling out if the tray is inverted.

Correct.

Wrong.

Carousel trays for 140 slides differ from those of eighty-slide capacity. Check the bottom of the tray to be sure that the *index hole* is opposite the *index notch*. If it is not, push the release latch in the direction shown, and revolve the bottom plate until the index notch and hole are aligned. Though the 140-slide capacity of the tray is often convenient, care must be exercised in using it: slide mounts must be in perfect condition, neither warped nor bent, and without frayed corners. Remount slides if necessary. Use of the 140-slide tray is not generally recommended.

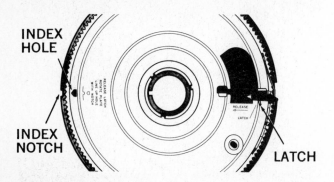

Projection Procedures

Examine the bottoms and sides as well as the tops of slide projectors. On the bottom of some you will find air-intake slots and access to the projection lamp and cord compartments.

Set up the projector. If a remote-control cord is to be used, attach it. Note provisions for focusing, changing image size, rejecting slides, and returning to previously shown slides.

On each projector you use, determine how to install the slide carrier or tray. Determine also how to remove a tray when the projector is turned off and turned on.

Turn on the projector lamp and adjust the projected image for size, position, and focus. Note provisions for adjusting the projector lamp setting (High for maximum brilliance or Low for maximum lamp life).

Show the slides. Practice using any special features provided on the projectors.

If possible, investigate the following less common techniques of using slide projection equipment:

• Showing slides in a tray-loading projector (such as the Kodak Carousel) without a tray.

• Using a stack loader accessory on a tray-loading machine. What are the advantages and disadvantages of this loading method?

• Using filmstrip projection attachments on a slide projector such as the Kodak Carousel.

• Several techniques are used to synchronize slide showings with compact cassettes carrying sound to accompany the slides. One popular method, such as used by the Wollensak recorder by 3M Company (shown below), carries slide change pulses on one of the stereo tracks, separate from the audio program. This arrangement permits inserting or changing pulses without affecting the program track.

Tips for Slide Use

To prolong the life and improve the appearance of your slides, store and handle them with care. To remove dust from slides, use a soft brush, an air bulb syringe, or a pressure can of dust remover. Avoid touching the film surface. To remove fingerprints, use film cleaner obtained from camera stores; follow directions carefully.

To provide a blank screen at the beginning and end of a slide series, make 2- by 2-inch squares of cardboard, or use blank or color-tinted slides. Use blanks also within slide series for transitions or when group discussion is programmed; a blank screen improves audience attention to the discussion.

When a slide series for formal lectures is prepared, repetition of slides is arranged by using duplicate slides at appropriate positions in the sequence.

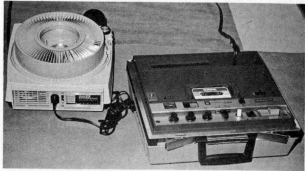

EXERCISE 58

SOUND-SLIDE VIEWERS AND PROJECTORS

Eastman Kodak Company.

The Reason for It

Ease of operation and simplicity, as well as effectiveness for individual viewing and for group presentations, accounts for the widespread popularity of automatic sound-slide viewers and projectors. A user does not have to operate the equipment, but only needs to know how to set up, cue, and start it. Sound-slide productions may have either or both production formats, in one format the audio cassette carries an audible pulse (beep) to tell an operator when to change slides; in the other, an inaudible pulse to change the slides on automatic equipment. Some programs have another pulse signal that will stop the automatic projection sequence until the operator restarts the tape.

Before You Start

Brown, Lewis, and Harcleroad, *AV Instruction*, Chaps. 9 and 10. This manual: Exercises 41, 46, 54, 57, 59; Correlated References, Section 20.

Purposes

1   To operate automatic sound-slide viewers and projectors

2   To put inaudible synchronizing ("synch") pulses on an audio tape to change slides in proper synchronization with the audio program

3M Company.

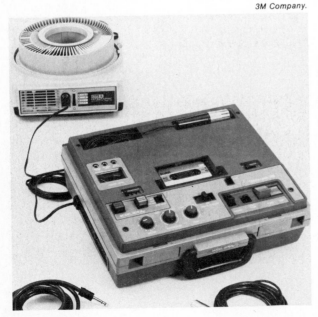

Required Equipment and Materials

1   Automatic sound-slide viewer or projector, or both

2   A commercially produced synchronized sound-slide presentation

3   An audio cassette tape deck with synch-record function or a sound-slide viewer with the record-pulse function

4   Blank audio compact cassette

Assignments

1   Load the automatic sound-slide viewer or sound-slide projector and tape player combination with a commercially produced program. Cue the program and the equipment and show the presentation. Any problems? If so, recue and start again. Determine source of any problems.

2   Select ten to fifteen slides, arrange them in a logical sequence, and record a short narration on a compact cassette to complement the slide story.

3   Record inaudible synch signals on the audio cassette recording to change the slides at the proper points in the narration.

4   Play back the slide program on the automatic equipment. Observe and evaluate the results.

5   Erase one or more of the synch pulses and rerecord them at different locations to alter the timing of slide changes. Replay the revised program and evaluate results.

Practical Pointers

•   Fortunately, synchronization pulses are becoming standardized at a frequency of 1000 Hertz (Hz) for the slide-change signal; for the pulse to stop the program, the signal is 150 Hz. Check to be sure that the program pulses are the same frequency used by the machine available.

- If corrections have to be made in the synch signals of a locally produced tape—if a signal occurs in the wrong place and must be moved, for example—be sure that the wrong signals are completely erased. (See equipment manual.)

- Practice until you are completely familiar with the particular make of equipment you are using. In some models special procedures are required for cueing, and the audio tape cannot be partially rewound after it is started. Other models permit partial winding without losing synchronization of the tape with the slides. Again, refer to the equipment-operating manuals.

- Kodak, Singer, and Bell and Howell manufacture slide-sound viewers that both project on the built-in screen for individuals and small groups and permit projection on a screen for larger groups, such as a medium-size class. An Eastman Kodak version of this feature is shown in the accompanying picture of the 460 Model that also permits recording and pulsing programs.

- It is relevant to mention that sound filmstrip machines also are made with automatic frame advance, using the same procedures as the sound-slide machines. One version, by Dukane Company, is pictured here.

Dukane Corporation.

EXERCISE 59

FILMSTRIP PROJECTORS

Dukane Corporation.

Singer Education Systems.

Viewlex.

The Reason for It

Either with or without sound, filmstrips are an economical medium for teaching with a fixed sequence of still pictures. They are widely available, moderate in cost, and convenient to use. The equipment for filmstrip projection for group viewing is simple to operate; inexpensive table-top viewers, with or without sound, permit filmstrips to be used for independent study.

Before You Start

Brown, Lewis, and Harcleroad: *AV Instruction*, Chap. 9; Reference Section 1. This manual: Correlated References, Section 20.

Purposes

1 To learn to protect silent and sound filmstrips

2 To learn to handle and care for filmstrips

3 To learn to set up, use, and maintain filmstrip projectors

Required Equipment and Materials

1 A 35mm filmstrip projector and a silent filmstrip

2 A screen or other matte white surface on which to project the filmstrip

3 A filmstrip with accompanying sound, and the appropriate audio equipment to play the cassette, tape, or disc provided

Assignments

1 With a filmstrip projector and a silent filmstrip (a) locate and identify the features and controls on the projector, and (b) practice the steps in setting up and operating it. Be able to demonstrate with confidence and accuracy.

2 Set up appropriate equipment to show filmstrips with sound. Demonstrate competence in synchronizing sound and picture when an audible signal for picture change is provided; then demonstrate the operation of synchronization by inaudible pulse signal.

3 Operate a silent viewer and an automatic, synchronized viewer with sound. In each assignment, demonstrate proper handling of both filmstrips and sound sources.

35mm Filmstrips

Single frame filmstrips are run downward through a vertical film channel in the projector or viewer. Since film is a relatively fragile substance, it can be finger-marked or scratched; therefore, handle filmstrips with care, hold them by the edges and by the blank film ahead of the titles (leader) and after the end title (tail).

Projection Procedures

The projector shown is typical of filmstrip machines. To operate:

1 Set it on a table or stand, and plug in the cord.

2 Turn on the lamp and elevate the front of the machine, using the adjustable front feet. Move the projector closer to or away from the screen to obtain an acceptable image size. Use the focus knob until the beam of light on the screen has sharp edges.

3 Pull up the film holder spool; with the filmstrip in the holding cup, bring the start end of the film over and in back of the holder spool. Insert the end into the film carrier slot, pushing it gently as far as it will go. Continue to push gently and at the same time turn the film advance knob counterclockwise until the film is pulled by the knob action.

131

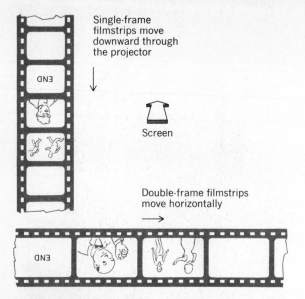

Single-frame filmstrips move downward through the projector

Screen

Double-frame filmstrips move horizontally

4 Advance the film until the first frame (focus, or title, or a picture) appears; adjust the framing lever until a complete frame is centered on the screen. Make any needed focus adjustment.

5 Show one frame after another by turning the advance knob briskly; to reverse the filmstrip, turn the advance knob clockwise.

6 When the showing is completed, remove the filmstrip and put it in its container. To make the filmstrip coil smaller, roll it into a tighter coil by holding edges only; *do not cinch it.*

7 Retract the lens and the front legs, coil the power cord, and store any spare parts or accessories. Store the projector in the carrying case, if provided, securing the latches.

Sound Filmstrips

Filmstrips with compact cassette sound may have audible frame-change cues, in which case you should use a standard filmstrip projector and a cassette player. Practice until you can properly synchronize the frames (titles and pictures) with the sound cues. If a filmstrip has cues for automatic projection, follow the procedure for automatic slide-changing. Thread the filmstrip through the projector or viewer and insert the cassette in the audio component. Synchronize the starting points on the filmstrip and the cassette. Prefocus the filmstrip and control audio levels as with any cassette equipment. Refer to instructions on the equipment or to the operator's manual, if necessary. Some models of filmstrip viewers can also be used as a projector for a group by a simple adjustment; other procedures are standard.

Many sound filmstrip viewer-projectors (below) have clear instructions within the operating portions of the case.

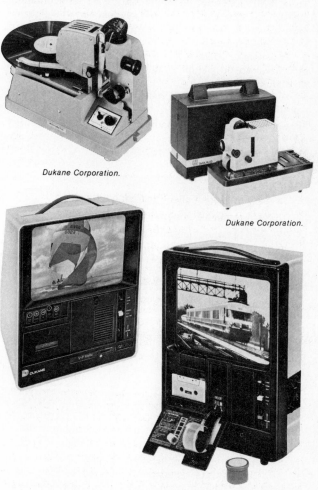

Dukane Corporation.

Dukane Corporation.

EXERCISE 60

OPAQUE PROJECTORS

The Reason for It

For efficient operation, the opaque projector requires a completely darkened room; nevertheless, it is valuable because of the abundance of available material that may be used with it. On the opaque projector, magazine or book illustrations, snapshots, newspaper clippings, and students' work may be shown. It also is useful to project drawings that are to be enlarged for tracing on the chalkboard or on a large sheet of paper attached to a wall or a bulletin board.

Before You Start

Brown, Lewis, and Harcleroad: *AV Instruction*, Chap. 9; Reference Section 1.

Purposes

1 To learn to operate opaque projectors

2 To learn to project opaque materials, including flat pictures, books, and small objects

3 To learn to use the opaque projector for enlarging sketches and drawings

Required Equipment and Materials

1 Opaque projector and screen

2 Materials for projection: magazines and illustrated books, standard and jumbo-size picture postcards, student-drawn pictures, small objects

Assignments

Learn to operate an opaque projector. Refer to the photographs in this exercise as you study the features and practice the operations described below.

1 *Set up the projector for safe operation.* When you place the projector on a table or wheeled cart, make certain that it grips the surface so that it cannot skid or be pushed off. A soft rubber cushion will help. If the power cord crosses a traveled floor, cover the cord with a rubber mat. Secure the cord to a leg of the table or cart.

2 *Learn how to tilt or elevate the projector.* The front feet of the projector may be on rods which can be extended individually to raise the front of the projector. Each leg is held in place by a lock or setscrew. Lower the projector to a level position for greater stability when it is not being used.

3 *Learn how to remove or adjust the lens.* Usually a setscrew or collar can be loosened to permit removal of the projection lens for replacement by a lens of different focal length, for cleaning of the rear lens surface, or to permit repositioning to provide clear focus when the screen is unusually close or far away. This is *not* the procedure for normal-focus adjustments, only for extreme adjustments. Do not

move the lens so far out that it is in danger of falling out of its mounting. With the lens collar retightened, use the focusing knob to focus the lens.

4 *Learn to load the platen,* which you may need to hold down while you position pictures for showing. After loading, release and raise the platen to its closed position where it will be held by springs. Some platens have a conveyer belt, moved by turning a crank, by which pictures can be fed into one side of the projector and out the other side. Does your projector have a similar device? Is it designed so that pictures can be changed with a minimum of light spill?

5 *Learn to place pictures on the lowered platen* so that they appear in the proper place and position on the screen. When you stand behind the projector and face the screen, will materials appear upside down to you as you place them on the platen? In what direction will the picture move on the screen as you move it left or right, up or down, on the platen?

6 *Show small objects or pictures from textbooks.* A piece of heat-resistant glass can be placed across the open book

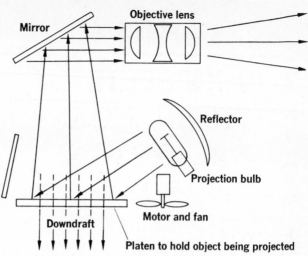

Mirror | Objective lens | Reflector | Projection bulb | Motor and fan | Downdraft | Platen to hold object being projected

to flatten it and thus improve the focus of the screen picture. Prolonged exposure in some opaque projectors may affect some pictures, especially colored ones; so avoid showing pages that are part of expensive or irreplaceable books.

7 *Practice removing the tray which contains the glass plate.* This plate holds book pages flat but should be removed for projection of small objects, such as rock specimens, which might scratch it. Do not show objects which may be damaged by heat. Any objects, especially those made of metal, may become hot to handle—use care!

8 *Practice using any arrow or pointing accessory that may be provided.*

9 *Note how pictures are cooled and sometimes held in place by the fan.* In most machines the fan and lamp are controlled by one switch. Some older opaque projectors are cooled by exhaust fans located high in the projector. These sometimes suck small pictures "up the chimney." In many newer opaque projectors the fan sucks air down through perforations in the platen. This suction holds pictures, even small ones, securely against the platen or platform. Pictures may also be held flat by metal frames, heat-resistant glass plates, small magnets, or by being mounted on heavy cardboard.

10 *Examine the inside of the projector.* Note the reflectors, which concentrate the light on the surface of the projected material. Inside the top of the projector a mirror reflects the projection image out through the lens. This may be a front-surface mirror, which can be easily scratched or damaged. Blow dust off the mirror; do not touch it, except with an *extremely* soft brush.

11 *Learn how to replace the projection lamp* (refer to Exercise 62). Disconnect the power cord, and allow the bulb to cool before attempting to remove it.

12 *Use the projector as an enlarger.* Trace a picture projected by the opaque projector onto a chalkboard or paper mounted on the wall.

An Application of Reflected Light

The above diagram explains how opaque projectors operate by reflected light, which accounts for their relatively weak image presentation. Used under properly darkened room conditions, they provide valuable service. Below is a recent design of an opaque projector, the VuLyte IV by the Charles Beseler Company.

Charles Beseler Company.

EXERCISE 61

PRINCIPLES OF 16MM FILM PROJECTORS

The Reason for It

Though on first acquaintance with a 16mm film projector you may think it a complex device, this machine is not really difficult to operate. The sound controls are like those of radios or record players, and the film transport and optical systems are similar to those of filmstrip projectors. Slot-loading and automatic threading projectors have further simplified operation procedures. Although different makes and models vary in appearance, their operation is basically similar.

Before You Start

Brown, Lewis, and Harcleroad: *AV Instruction*, Chap. 11; Reference Section 1. This manual: Exercises 62, 68; Correlated References, Section 18.

Purposes

1 To provide background information about principles of 16mm film projection that will be helpful as you learn to operate various types of projectors

2 To introduce the steps in showing a film common to most manual-threading, slot-loading, and automatic-threading projectors

Required Equipment and Materials

1 A 16mm manually threaded sound projector

2 A slot-loaded 16mm projector

3 A 16mm projector with automatic threading

4 A reel of 16mm sound film (a short film, under 50 feet, is recommended) and a take-up (empty) reel

5 A gate brush or other recommended cleaning tool

6 A projection screen

7 An operating manual for each projector used

Assignments

1 Study the three systems of the generalized 16mm motion picture projector on the next page. Learn the function of each component listed. Locate and identify the same components on a 16mm manually threaded projector. You may find that some machines do not have features described in the list; checkmark these. You may also find components on some machines that do not appear on the list; refer to the manufacturer's manual to identify these items, and add them to the list.

2 After reading the remainder of this exercise, set up, thread, operate, and put away a manually threaded machine.

Refer to Exercise 62. If instructions for operating the machine you have are not included there, examine the body of the projector or the inside of the lid, where operating instructions may be printed, or refer to the operator's manual for the machine.

3 After operating a manual machine, repeat assignments 1 and 2 above, using slot-loading and automatic-threading machines. Again refer to Exercise 62 and/or the operator's manual for each machine.

Characteristics of 16mm Film

Most 16mm film is used for sound-accompanied projection. Examine a piece of this film. Occasionally, one might use silent 16mm film; for this reason it is described here.

Sound and silent films are projected at different speeds; the projector must be switched to the proper speed for the type of film used. Sprocket holes in the film mesh with teeth on the sprocket wheels in the projector which pull the film through the machine. What will happen if the sprocket holes are torn or otherwise damaged?

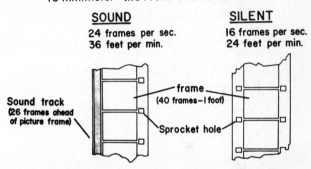

16 millimeter MOTION PICTURE FILM

SOUND
24 frames per sec.
36 feet per min.

SILENT
16 frames per sec.
24 feet per min.

frame (40 frames—1 foot)

Sound track (26 frames ahead of picture frame)

Sprocket hole

Sound tracks may be optical or magnetic. Three types of optical tracks are shown below; any of them can be used on any 16mm optical sound projector. Some 16mm projectors are equipped for magnetic sound tracks, which reproduce sound as do audio tapes.

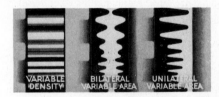

VARIABLE DENSITY   BILATERAL VARIABLE AREA   UNILATERAL VARIABLE AREA

Sixteen millimeter motion picture film is mounted on reels of several sizes. Conventionally, the 400-foot reel is considered "one reel"; other sizes are increments of this size. The chart shows the relationship of film footage, reel size, and showing time.

16MM REEL SIZE—RUNNING TIME AT SOUND SPEED		
Feet of Film	**Reel**	**Time in Minutes**
400	1	11
800	2	22
1200	3	33
1600	4	44
2000	5	55

A GENERALIZED 16MM SOUND MOTION PICTURE FILM PROJECTOR

All sound motion picture film projectors must have three basic systems: one for reproducing sound, another for projecting the image carried on the film, and a third for transporting the film through the projector. Important parts of each system and their functions are listed below. Study each system and attempt to recognize its components on actual projectors (Exercise 58).

Sound System Components

- A *power cord* supplies AC power.

- *Amplifier controls* turn the amplifier on and off and adjust volume and tone. In some projectors functions may be combined, e.g., the tone control knob may also incorporate the amplifier off-on switch.

- A *speaker* may be mounted in the projector case or may be a separate unit. If separate, a *speaker cord* is used to connect the speaker to the amplifier in the projector.

- For optical sound, light from the *exciter lamp* passes through the film sound track to the *sound drum*, where a photoelectric cell converts the pulsing light into electrical impulses.

- For magnetic sound, *magnetic heads*, like those on tape recorders, play back or record sound on a magnetic stripe on the film.

Projection System Components

- A *projection lamp* and a *lamp switch* (which in some machines may be incorporated in the motor switch)

- A *lens*, which is adjusted to focus the image

- An *elevator knob*, which raises the projector to center the light beam on the screen.

- A *framer*, which adjusts the projector aperture so that only one complete picture is seen at a time on the screen

Film Transport System Components

- A *feed reel* and a *take-up reel* are supported on *reel arms*. (In some machines, drive belts to rotate the reels must be attached when setting up the projector.)

- A *motor switch* controls power for the film transport system motor.

- Toothed *sprocket wheels* (usually two or more) mesh with sprocket holes in the film to drive it smoothly through the machine.

- *Guide rollers* prevent the film from rubbing against the projector case, and *snubbers* cushion the film against strain.

- A *film channel* incorporates the aperture in the lamp housing, where the light passes through the film and the *film gate* on the lens housing.

- A *hand operation knob*, when turned, moves the film through the projector to check threading.

Differences Between Semi-Automatic (Slot-Loading) and Automatically Loading 16mm Projectors

On *manually threaded machines,* the film must be put into place in the entire film path. Using the fingers, the sprocket holes must be engaged with the sprocket teeth and latched into place. The loops must be properly formed above and below the aperture plate area; the film must be laid into the gate area, and then guided around the sound components, past the guide roller(s) and out to the take-up reel. (See drawing, above.)

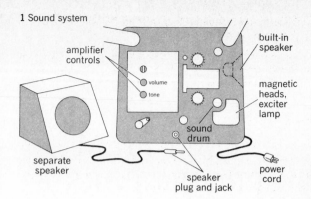

1 Sound system

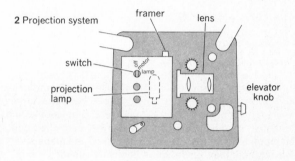

2 Projection system

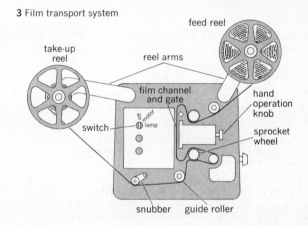

3 Film transport system

Though many slot-loading and automatic machines are now in use, manual projectors are still available. We therefore advise operators to understand them and to be prepared to operate them.

On *slot-loading machines,* the film is loosely laid by hand into a slot that indicates the film path through the three systems shown above—film transport, projection, and sound. A single Thread level readies rollers and guides that will direct the film exactly through the entire film path, form loops, engage sprockets with sprocket holes, and run the film around the sound drum components with proper tension, where it is ready to attach to the take-up reel. Slot-loading machines can be quickly unthreaded when the Run lever is released.

Fully *automatic threading machines,* also simple to operate, require cutting the end of the leader perfectly square. This is done in a trimming knife which is mounted on the projector. After trimming, the leader is fed into a slot or guide at the top of the projector; it will then run rapidly through and out the lower back of the machine. After a slight tug on the film to release the automatic threading guide system, attach the leader on the take-up reel. At this point, it is ready for projection.

TYPICAL OPERATING PROCEDURES— ALL 16MM MACHINES

Except as noted, the following sequence of basic steps in 16mm motion picture projection is applicable to all typical projectors, manual, slot-loading, and automatically threading.

Setting Up

1 Put the projector case on a stand, and open the case.

2 If the speaker is mounted in the cover or is in a separate case, attach the speaker cable at the projector. Place a movable speaker where all will be able to hear.

3 Plug the power cord into a wall receptacle. Secure power and speaker cords to projector stand legs.

4 Turn on the amplifier.

5 Latch the reel arms into the operating position.

6 If there is a film speed selector, set it for the film used. If there is a reverse control, set it at Forward.

7 Turn on the motor and lamp. Elevate the projector to center the light on the screen and move the projector toward or away from the screen until the light fills the desired area.

8 Focus the light beam by turning the lens barrel (or the focus knob) until the edges of the lighted area are sharp. Examine the edges for evidence of dirt in the film-channel aperture.

9 Turn off the projector. Open the film channel, and with a gate brush clean all surfaces over which film passes. *Note:* It is recommended that this be done each time a film is threaded.

10 Check the sound system. Is the amplifier turned on? On some machines the exciter lamp comes on only when the projector is started; on others it lights when the amplifier is turned on.

11 Push the reel of film firmly on the spindle. *Note:* The reel lock on some machines operates simply by pushing the reel on the spindle as far as it will go; on others a snap-down catch is used.

12 Check to make sure that the film is properly wound on the supply reel. As the film comes off the front of the reel, the image should be head down, and the sound track should be on the side nearest the projector.

13 On the take-up arm spindle, attach an empty reel, as large as or larger than the supply reel.

Threading: Manual Machines

Refer to threading instructions in Exercise 62 or to a chart on the projector, or to the manufacturer's manual for specific types and models of machines. On each machine threaded manually, careful attention should be given to the following important points:

1 Teeth on each sprocket wheel must fit through holes in the film. The film must lie smoothly in the film channel.

2 Establish loops above and below the film channel according to lines or embossings on the projector housing (or by a special guide on some machines). The lower loop must be made accurately to ensure proper synchronization of sound and picture. If a loop is lost during a showing, the picture will vibrate on the screen; to correct, stop the projector and re-form the loop, or use a loop-setting lever, if provided, and follow manufacturer's instructions.

3 The film must be under tension as it passes around the sound drum. If film is threaded too loosely here, the sound will be distorted. Some projectors feature special devices to ensure proper tension; on others extra care must be taken in threading around the sound drum to prevent distorted sound.

Threading: Slot-Loading Machines

1 Set the motor function switch to Off. Set any control lever to Load. Since each brand of projector has slightly different arrangements and terminology for threading procedures, turn to the next Exercise for details for each machine.

2 Holding the end of the film, draw the film into the threading slot following any guide lines or other indicators. Draw the film through the slot and attach the end to the take-up reel; rotate the reel clockwise a full turn or two to catch the film firmly on the reel.

3 Engage the control lever to the Project or Run position or Forward. Do not switch the machine to On to move the mechanism.

4 First, if provided, use the manual or inching knob (usually on the front of the machine) to advance the film through the mechanism to be sure the film is engaged and moving smoothly.

5 Then turn the projector *and* lamp on or set on Forward or Run. Determine that the film is running smoothly; then, promptly and only if necessary,

a Adjust focus: picture should be sharp and clear.

b Adjust sound: set volume level and tone.

c Adjust frame line, if necessary using the framing lever.

d Make fine adjustments to volume and tone.

e Stop the projector, then run it in Reverse back to the beginning to start the showing.

Threading: Automatic Machines

Automatic threading greatly simplifies film loading procedures. When the film is inserted in the film threading slot and the machine is turned on, the projector mechanisms automatically perform all remaining threading operations, except attaching the film end to the take up reel. Set up the projector as described previously for a manual-threading machine. Then thread the projector as follows:

1 Be sure projector controls are properly set for loading (refer to the next exercise, instructions on the projector, or manufacturer's manuals).

2 Trim the film leader. Insert approximately 2 inches of leader into the cutter, and clip off the end of the film. *Note:* The first 3 feet of leader must not have any tape, torn sprocket holes, crimps, misaligned splices, or other irregularities.

3 Turn the projector on, and insert the end of the leader into the threading slot. When two feet or so of leader have passed out the rear of the machine, stop the projector. Complete threading by attaching the film end to the take-up reel.

Showing the Film

Whenever possible, run a few feet of film to check both picture and sound quality. Stop the machine. Then reverse the projector, and run the film back to the starting point ready for the showing.

The following procedures apply generally to all typical classroom motion picture projectors:

1 Set sound-volume control at about one-third or one-half of full volume. Set tone at the midpoint. Turn off room lights.

2 Start the showing by the following steps:

a Turn on the motor.

b Turn on the lamp as the title nears the aperture.

c Adjust focus the instant an image appears on the screen.

d Adjust volume, then tone.

e Correct the framing, if necessary.

3 During the entire showing, stay with the projector to:

a Correct focus and adjust sound, if necessary.

b Shut off the machine immediately in case of faulty operation.

4 To close the showing:

a As the end title fades, turn off the lamp. As the sound fades, or ends, turn the volume down to *zero*.

b When all the film is on the take-up reel, turn off the motor. Turn off the amplifier if no more films are to be shown.

Rewinding

To rewind the film, reels must turn opposite to the direction they turned when projecting. Thus, on projectors threading on the right side, the reels turn counterclockwise when rewinding. Refer to instructions for specific machines in the next exercise, on the projector itself, or in the manufacturer's manual.

Putting Away the Projector

1 Coil all cords, and place them in the spaces provided.

2 Dismantle the machine, and put the reel arms, cables, and spare reel in the positions provided for storage.

3 Retract the elevating mechanism.

4 Make certain all switches are off and all levers are in their normal forward operating positions (not in reverse or rewind).

5 Close and lock the lids firmly.

A PORTFOLIO OF HINTS: 16MM PROJECTORS

Sprocket holes in film should engage sprocket teeth.

oops!

Use a take-up reel of adequate size.

Trim film.

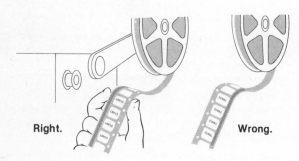

Right.   Wrong.

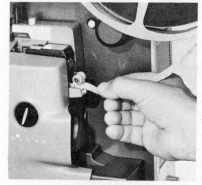

Place film in threading slot.

A Useful Chart:
Lens/Distance/Screen-Size Relationships

Most 16mm film projectors are supplied with lenses of 2-inch focal length since they provide an image of appropriate size for viewing in many situations. In carrels or auditoriums, however, because projection distance and image size requirements present special problems, lenses of other focal lengths may be more appropriate. The chart at the right shows relationships of focal lengths, distances, and image sizes.

Lens focal length, in.	Width of screen								
	40 in.	50 in.	60 in.	70 in.	84 in.	8 ft.	9 ft.	10 ft.	12 ft.
	Projector-to-screen distance, ft.								
1	9	11	13	15	18	21	24	26	32
1½	13	17	21	24	30	34	38	42	50
2	18	23	27	32	39	44	50	55	66
2½	22	27	33	38	46	53	59	66	79
3	26	33	40	46	55	63	71	79	95

EXERCISE 62

16MM MOTION PICTURE PROJECTORS

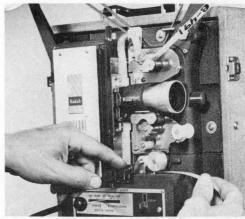

The Reason for It

Since there is only one way to become a skillful projectionist —to practice—this exercise provides basic steps to guide you in learning to operate typical projectors used in instructional situations. As you use the machines available to you, work with each one until you can operate it with precision and assurance.

Before You Start

Brown, Lewis, and Harcleroad: *AV Instruction*, Reference Section 1. This manual: Exercises 55, 61; Correlated References, Section 18.

Purpose

To develop acceptable skill in operating several typical 16mm projectors

Required Equipment

16mm sound projectors, with essential accessories (empty take-up reel, practice film on a reel, power cord), and a copy of the operator's manual for each machine

Assignment

Practice with each machine assigned using the step-by-step procedures that follow. Test yourself, using the laboratory checksheet at the end of this manual.

KODAK PAGEANT PROJECTOR— MANUALLY THREADED

Setting Up

Follow the general procedure outlined in Exercise 57. On the Pageant projector, remove the case front and set up the speaker. Raise the reel arms; install the take-up arm belt. Be

sure the rewind tab is in the vertical position before turning the machine on to aim the lens, adjust picture size, and set the speed selector.

Threading

1 Open the feed and take-up sprocket clamps and the gate. Push forward on the gate tab until it latches.

2 Turn the hand threading knob until the white line on the knob is toward you. Engage the film in the upper feed sprocket, and set the top loop by aligning the film with the red dot on the rewind tab. Close the gate.

If the machine does not thread properly, shut off the motor immediately, unthread manually, and check the film condition and recheck your threading procedure.

3 Thread the film under the loop-setting roller; the film should just touch the roller. Thread over the sound drum pressure roller and under the sound drum; then thread behind the damping roller, over the sprocket, and out. Be sure the film sprocket holes engage the sprocket teeth.

4 Press down on the loop setting roller as far as it will go, then release it; this action forms both top and bottom loops correctly.

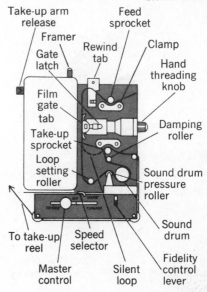

139

5 Feed around the snubber roller and under the two rollers on the bottom of the master control cover. Attach the film end to the take-up reel.

Showing the Film

Refer to instructions in Exercise 57. A fidelity control lever adjusts sound for optimum quality. To operate the projector in reverse, move the master control lever to *reverse-motor* and, if desired, to *reverse-lamp*.

After the Showing

• To rewind: Attach the film from the take-up reel to the empty feed reel. Move the master control to *rewind*; set the speed selector at *sound*. Lower the rewind tab to its horizontal position.

• To put away the projector: Remove the take-up arm drive belt, raise the arm slightly, push the arm release button, and lower the arm to its storage position. Swing the supply arm down, lower the projector, stow speaker and power cords, and replace the cover.

FOUR SLOT-LOADING 16MM PROJECTORS

Setting Up

Follow the general procedures for setting up in Exercise 61.

Bell & Howell.

EIKI International.

Kalart-Victor.

Singer.

Threading

1 Set the motor function switch to Off. Set the control lever to Load. Since each brand of projector has slightly different arrangements and terminology for threading procedures, separate instructions follow for each of the four slot-loading machines.

• *Bell and Howell (Model 2580)* Push the Load lever down to position 1 (Load) during threading. After threading move lever up to position 3.

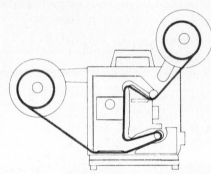

EIKI International.

• *Eiki (Model SL)* Turn function knob to Off during threading. After threading, turn knob to first detent before (above) positions 1 and 3, and rotate the inching knob.

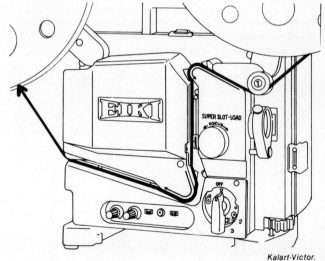

Kalart-Victor.

140

- *Kalart-Victor (Model 80)* The function knob should be off during threading. Unlatch and raise the Auto-Set lever as far as possible (top of machine, above entry to slot). After threading, lower the Auto-Set lever to the latched position.

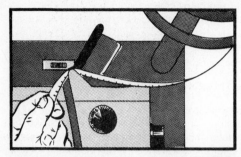

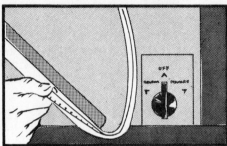

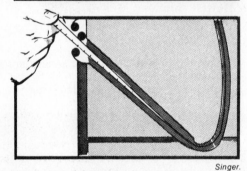

Singer.

- *Singer (Models 2100/2200)* Pull down Master Control lever (on rear of projector) to Load position. After threading, raise lever to Project.

2   Holding the end of the film, draw the film into the threading slot following any guidelines or other indicators. Draw the film through the slot and attach the end to the take-up reel; rotate the reel clockwise a full turn or two to catch the film firmly on the reel.

If the machine does not thread properly shut off the motor immediately, unthread manually, and check the film condition and recheck your threading procedure.

Singer.

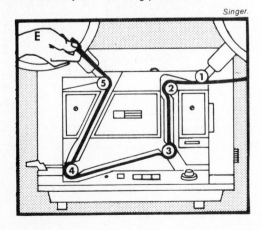

3   Engage the control lever to the Project or Run position or Forward. Do not switch the machine to On to move the mechanism.

4   First, if provided, use the manual or inching knob (usually on the front of the machine) to advance the film through the mechanism to be sure the film is engaged and moving smoothly.

5   Then turn the projector *and* lamp on or set on Forward or Run; determine that the film is running smoothly; then, promptly and only if necessary:

   a   Adjust focus: the picture should be sharp and clear.

   b   Adjust sound: set volume level and tone.

   c   Adjust frame line, if necessary using the framing lever.

   d   Make fine adjustments to volume and tone.

   e   Stop the projector, then run it in Reverse back to the beginning ready to start the showing.

Showing the Film

1   Turn off the room lights

2   Turn the projector On.

3   Make any needed adjustments in this order:

   a   Focus

   b   Volume and tone

   c   Frame line

4   During the entire showing, stay with the projector to:

   a   Correct focus and adjust sound as necessary.

   b   Shut off the machine immediately in case of faulty operation.

   To close the showing:

   a   As the end title faces, turn off the lamp.

   b   As the sound fades or ends, turn the volume down to zero.

   c   When all the film is on the take-up reel, turn off the motor.

   d   Turn off the amplifier if no more films are to be shown.

Rewinding

There are two modes in which films can be rewound on slot-loading projectors: directly from the take-up reel to the feed reel across the top of the machine, or through the slot path (except on the Bell & Howell). However, the direct reel-to-reel method is recommended to avoid unnecessary wear on both the projector mechanism and the film.

Before changing the direction of the film movement on any machine, stop the machine completely at the Off position. *Reel-to-reel direct rewinding:*

1   Allow the film to run completely onto the take-up reel. Turn the machine off.

2   Draw the film directly across the top of the projector from the feed reel to the take-up reel and attach the leader securely.

3   Operate the Rewind control and allow the film to run completely to the take-up reel on the right.

Rewinding Through the Slot Path

Though this procedure is not recommended, film can be rewound through the slot path by these steps (except on Bell and Howell):

1   Stop the machine before the film trailer becomes detached from the take-up reel.

2 Release the master control lever. The film will run backward through the slot path to the feed reel.

Putting Away the Projector

1 Coil all cords and place them in the spaces provided.

2 Dismantle the machine, lower the reel arms, coil and store cables and the spare reel in the positions provided for them.

3 Retract the front elevating mechanism.

4 Be certain all switches are off and all levers are off or in their Forward operating positions (not in Reverse or Rewind).

5 Close and lock all lids firmly.

SPECIAL FEATURES OF SLOT-LOADING PROJECTORS

Each model of projector may have different features that affect operation. Such features are the result of options available to meet users' special needs—some make the difference between economy and deluxe models. Here are some features worthy of special note; mentioning them also serves to reinforce the recommendation that operators have a manual of instructions for the model being used.

Bell & Howell Filmosound (Model 2580—Slot-threading)

• Restoring lower loop: Depress loop-restorer lever below gate area.

• In case of faulty threading, with projector off, set load lever up to 3; be sure film is taut throughout path from reel to reel; move load lever down to 1.

• Rewinding: On Bell & Howell projectors, the rear (take-up) arm is horizontal when running film either forward or in reverse. When rewinding, the rear arm must be unlatched and raised to the vertical position and latched. The motor switch is set on Reverse for rewinding.

• Cleaning the film path: With the projector off and unplugged, and the load lever down to 1 (load position), open the lamp house cover on the machine side near the operator. Use a Bell & Howell cleaning tool or a gate brush to clean all surfaces over which film passes. Give special attention to the aperture, or gate, area where light passes through the film. The heated aperture area on *any* projector may accumulate dirt and hardened emulsion that can scratch film.

(An option of Model 2582 only is still projection. A Still/Run lever on the lower front of the projector should remain on Run for all circumstances except to show a single, still picture; move the lever from Run to Still only when the machine is running.)

EIKI (Model SL—Super-Slot Load)

• Rewinding: Use the Rewind position on the function lever. The rewind path is typical of slot loaders.

• Lost loop: In case of a lost lower loop, an automatic loop restorer functions. A lost upper loop requires stopping the machine and resetting the master control back to Off, ensuring that the film is correctly in the film path, and restarting the machine to run position 3.

• Cleaning the film path: The film gate and aperture plate can be removed for cleaning. Use a gate brush to remove any hardened emulsion or loose dirt.

Kalart-Victor (Series 90—Easy-Load)

• Loss of loops: In case of a lost lower loop, a built-in loop restorer will operate. For a lost upper loop, however, a film safety switch will operate to stop the machine. Turn the operating knob to Off. Raise and lower the Auto-Set lever (top of machine). Turn on. Repeat sequence, if necessary.

• Cleaning the film path: Raise the Auto-Set lever and lift the lamp house cover. Use an aperture brush to clean the gate area and film path thoroughly.

Singer (Model 2100/2200—Instaload)

• Loss of loops: In case of loss of lower loop, an automatic loop restorer functions. In case of loss of upper loop, stop the projector, move the master control lever to Load, and turn the take-up reel to advance the damaged film area. Move the master control to Project, check film motion with the Inching Knob, then operate Forward button.

• Cleaning the film path: Remove the lens cover plate which carries the shoe and uncover the aperture and gate area. Carefully clean all the polished surfaces of the film path with an aperture brush. Use care not to scratch the polished surfaces.

OPERATING AUTOMATICALLY THREADED PROJECTORS

Bell & Howell Autoload Models

Set up autoloading models following the method used with other types of machines. Thread as follows:

1 Inspect the first several feet of film; it must be undamaged and free of tape or obstructions. Trim the end of the leader in the film cutter on the front of the machine.

Bell & Howell.

2 Turn the control switch to Forward.

3 Push the autoload lever forward until it locks in position.

4 Insert the film in the threading slot until it engages the upper sprocket.

5 When about 2 feet of film have come out the back of the machine, pull lightly on the loose end of the film until a click is heard, then stop the projector and attach the film to the take-up reel.

Procedures for operating the projector are the same as those for slot-load projectors. To rewind, follow the same procedure as for other Bell & Howell projectors.

To unthread the film manually, open the threading mechanism door and remove the exciter lamp cover. Turn the supply reel to provide slack in the film. Starting at the upper sprocket, ease the film out of the film path.

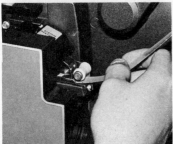

EIKI Self-Threading Machines (N.T. Models)

Set up the self-threading models as is done with other machines. After prefocusing and mounting feed and take-up reels, turn the machine off. Check that the rewind lever, top right of machine near base of the take-up arm, is up.

EIKI International.

1 Inspect the first two feet of film coming off the feed reel for any damage or adhesive tape. Insert the end of clean, undamaged film in the film cutter on top of amplifier (below the lens) and trim the end cleanly. (A)

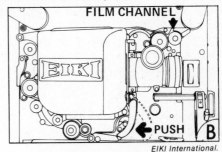

EIKI International.

2 Turn the function knob to Forward.

3 Find the Auto-Thread control lever below the gate and amplifier; push it to Self-Thread position. (B)

4 Insert cut end of film leader into the film channel, top of machine. Film will thread automatically. When about two feet of film emerges from the back of the machine, a gentle tug on the film will release the automatic threading guides; turn the control knob off. (C)

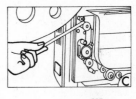

EIKI International.

5 Attach the film to the take-up reel.

6 Turn the control lever to Forward/Lamp. Make standard checks of focus, framing, volume, and tone. The machine features Still Picture and Reverse projection.

To rewind, attach the film tail to the supply reel directly across the top of the machine. Move the rewind lever (top front of projector) to the Down position. Turn the function knob left of Off to the Rewind position. When this is done and the machine is Off, return the rewind lever to the Up position.

Good Practices

Under *good practices* appear a number of items that represent safety precautions, proper care of materials and equipment, and techniques that lead to smooth and effective presentations. Moreover, many of the items are matters of courtesy to the next user of the machine.

• Always clean the film gate before every threading. This simple precaution will protect film from damage.

• Be sure to push reels on the spindles until they latch firmly. If the reel spindle has a snap-down catch, make certain the catch is latched to secure the reel.

• Always select a take-up reel at least as large as the feed reel carrying the supply of film.

• Check carefully the engagement of the film sprocket holes with sprocket teeth.

• To best judge sound quality and volume, step a few feet away from the machine; projector noise may cause you to misjudge the sound.

• If a projector has a Hi-Lo switch on the lamp circuit, and if the low setting gives a satisfactory picture, use it to prolong lamp life.

• Never leave a projector unattended during a showing. If the picture on the screen starts to bounce or if the film passes the aperture without stopping at each frame, or chatters at some point on the path of travel, *stop* the projector immediately; then look for the trouble.

• When rewinding, touch the spinning reel with care. Fingers should touch only the sides of the reel rims and should point in the direction of reel rotation to prevent injuring yourself.

• If the film breaks, wind it overlapped on the take-up reel, mark the break with a slip of paper, and continue with the showing. *Never, under any circumstances*, make temporary repairs with gummed tape, paper clips, pins, or other materials. Film must be spliced with special cement only, and a film-splicing machine must be used. Mark a slip of paper indicating the type and location of film damage, return the film

to the can without rewinding, and leave the slip in the can with the edge extending; this marker will signal the film inspector that damage is being reported. *Note:* It is not a crime to break a film. But damaged film must be reported as a courtesy to the film owner and the next user.

• If a microphone input is provided on your sound projector, practice using it as a public address system. Try narrating the commentary instead of using the sound track.

• Always replace all parts in the projector case when putting the machine away.

• Always report faulty equipment to the office responsible for maintenance.

Do You Know...?

Check your knowledge and skill in operating 16mm sound motion picture projectors by considering the following questions:

1 What is the difference between "reverse" and "rewind"?

2 Why is it not practical to rewind films just by reversing the projector?

3 What would happen if the frames of the film were shot faster than sound speed and projected at sound speed? If the frames of film were photographed much more slowly than sound speed and projected at sound speed?

4 What is the running time of 400 feet of 16mm sound film? Of 1,200 feet?

5 What should you do if your film breaks in the middle of a showing?

EXERCISE 63

8MM FILM PROJECTORS AND VIEWERS

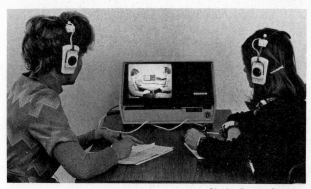

The Reason for It

Super 8 film has given some impetus to development and use of projectors that provide for both sound and silent instructional programs. The film may be mounted in cartridges or on reels. One type of Super 8 machine will send images and sound directly into a TV receiver or system.

Before You Start

Brown, Lewis, and Harcleroad: *AV Instruction*, Chap. 11; Reference Section 1. This manual: Correlated References, Section 18.

Purposes

1 To determine advantages and limitations of various configurations of 8mm motion picture projectors

2 To develop skill in handling 8mm equipment

Required Equipment and Materials

1 Several 8mm projectors of different brands and configurations, as available; silent and sound, regular 8mm and Super 8mm, reel-to-reel or cartridge load, manual and self-threading

2 Reels and cartridges containing 8mm film

3 An 8mm sound recording projector, if available, with film carrying magnetic stripe for sound, or a standard cassette or open-reel audio tape recorder

Assignments

1 Check all projectors available to you. Determine whether each is for standard 8mm film or Super 8mm film, or whether your machine will project both sizes.

2 Examine short pieces of standard 8mm film and Super 8mm film. What are the basic differences between the two types of film? Estimate the advantages of Super 8mm that might make it superior to standard 8.

Special-Purpose Super 8mm Viewer

The Beseler Cue/See combines compact cassettes and Technicolor Super 8 Cartridges for still or motion projection with sound. Pulses to advance the film are put on the audio cassette with accompanying sound for the film. Film can be advanced frame-by-frame, and may be cued to show still pictures like a filmstrip, or slow motion at any speed, or normal sound speed. These characteristics of the Cue/See viewer should be considered for interactive student lessons, for short instructional sequences, and for flexibility of presentations, including simple animation.

WORKSHEET 63

Name _____

Course _____ Date _____

YOU MAY COPY THIS FORM TO COMPLETE THE EXERCISE

Projector Name and Model

FILM TRANSPORT	Super 8						
	Standard 8						
	Both Super 8 & standard						
	Cartridge load						
	Reel-to-reel						
THREADING	Totally self-threading						
	Partially self-threading						
	Manual threading						
PROJECTOR CONTROLS	Motor switch						
	Lamp switch						
	Focus						
	Frame						
	Elevation						
	Reverse						
	Single frame						
	Speed control						
	Rewind						
SOUND	Optical						
	Magnetic						
	Cassette or ¼-inch tape						
	Volume control						
	Tone control						
	Loudspeaker						
	Earphone jack						
	Microphone input						
PROJECTION SURFACES	Wall screen						
	Screen in case lid						
	Rear-screen viewing box						
	Self-contained rear screen						
MAXIMUM REEL SIZE							
OTHER FEATURES	Choice of lenses						
	Cost						

Rate the projectors you have studied in order of preference for:

Individual viewing   1 _____ 2 _____ 3 _____

Small-group viewing   1 _____ 2 _____ 3 _____

Classroom viewing   1 _____ 2 _____ 3 _____

EXERCISE 64

TELEVISION RECEIVERS

The Reason for It

Learning from and enjoying television programs from any source depends much upon the quality of picture and sound received. Teachers should recognize inferior picture quality caused by improper receiver adjustment and should be able to make necessary changes in the appropriate control settings with speed and accuracy on both black-and-white and color receivers.

Before You Start

Brown, Lewis, and Harcleroad: *AV Instruction*, Chap. 11; Reference Section 1. This manual: Exercises 49, 66, 68; Correlated References, Section 21.

Purposes

1 To learn to recognize and correct faulty television pictures and sound by operation of appropriate receiver controls

2 To be able to determine by examining your TV receiver what accessories may be used with it

Required Equipment and Materials

1 A television receiver in satisfactory operating condition, properly connected with an adequate antenna or closed-circuit distribution system

2 Operating manual or instruction sheets for your receiver

3 A video player and an external speaker, if the receiver permits connection of external accessories

Assignments

1 Turn on your receiver and tune to an off-the-air broadcast channel. The following adjustment of controls will give you a feel for their effect and a knowledge of the picture problems misadjustment will produce.

a Turn off automatic fine tuning, if provided. Turn the fine tuning control through its full range. Note that the picture becomes very poor or goes away at the extremes of the range. Note the point at which the picture is the clearest. Is the sound most intelligible at this point or at some other point?

b Similarly, turn the brightness through its range. At what point does the picture go out of focus or lose detail in bright areas?

c Now, turn the contrast control through its full range. At what point does the picture lose clarity in dark or shadow areas?

d If yours is a color set, turn off the automatic color control, if provided, and then run the tint and color controls

through their ranges. Which points give just enough color and realistic flesh tones?

e Locate the horizontal- and vertical-hold controls. One at a time, rotate each control clockwise and counterclockwise until the picture becomes unstable. Note the different appearance of the unstable horizontal and unstable vertical pictures. Then return the controls to the middle of their stable range.

2 If your set has "monitor" functions, do the following:

a Connect an external speaker. Adjust any necessary switches and readjust volume and tone controls. What new settings are needed? Is sound quality improved?

b Connect a video player to the set. Adjust switches and play back a tape. What picture and sound readjustments were required when changing from off-the-air reception to video tape playback?

c If your player has an R-F (radio-frequency) output, connect this to the receiver's antenna input and tune as for broadcast programs.

Setting Up Receivers

• Place receiver so that extraneous light does not reflect from the receiver screen.

• Window darkening may be used, when required, but leave on the room lights that do not create glare on the receiver screen.

• Receiver height should be appropriate for comfortable viewing. When viewers are seated in chairs, the lower edge of the screen should be about 48 inches above the floor. When carpets are available, viewers may prefer to sit on the floor with receiver height lowered appropriately.

• Turn on the receiver well ahead of the viewing schedule and adjust for optimum picture and sound quality. Follow these steps:

1 Select channel and adjust fine tuning.

2 Adjust brightness and contrast.

3 Adjust sound volume and tone, if tone control is provided.

4 Adjust color and tint, if provided.

5 To limit distractions in the room until viewing time approaches, turn the receiver to Stand-By or to an unused input if these are provided. If not, turn down volume and brightness, making proper readjustments at viewing time.

Special Features on Receivers

Many classroom receivers have special features not provided on home sets. Inspect your receiver: Does it have a video input which permits attachment of a cable from a TV camera or a video tape recorder for playing tapes? Can an audio or video tape recorder be connected to record programs received by the television set? Copyright laws restrict what you may legally record. What is the policy of your school district on reproduction of copyrighted materials?

Check your receiver manual for any information not clear from an examination of the front and back panels of the receiver.

TELEVISION RECEIVER ADJUSTMENTS

Picture Courtesy of General Electric.

Picture Courtesy of General Electric.

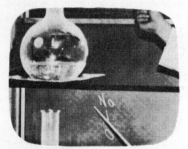

Picture Courtesy of General Electric.

Fine Tuning

Careful adjustment of the fine-tuning knob, usually behind the channel-selector knob, may be essential to obtain a clear picture.

Brightness

The brightness control knob adjusts the amount of light coming from the picture tube. In brightly lighted rooms, turn the brightness up (turn the knob clockwise); then adjust the picture contrast.

Contrast

Contrast is the ration of black to white in the picture. The illustration shows the picture with too much contrast. You may have to correlate adjustment of brightness and contrast to obtain the best possible picture.

Picture Courtesy of General Electric.

Color

Color receivers have only two principal adjustments beyond those for black and white:

Tint: Adjust hue to provide the most natural flesh tones (neither too green nor too blue).

Color: Adjust to provide an intensity of color to suit personal taste, always seeking the most natural effect. With saturation adjustment too high, colors are unnaturally bright and the picture may appear grainy (noise in the picture); with saturation too low, colors tend to be weak or to disappear.

Vertical Hold

Sometimes the picture rolls and the image seems to move upward or downward. Turn the vertical control knob to the left or right until the picture stabilizes on the screen.

Horizontal Hold

Sometimes the picture seems to become a series of narrow diagonal segments. The horizontal hold knob should be turned to the left or right until the picture is stable on the screen.

TELEVISION TRANSMISSION FAULTS

Picture Courtesy of General Electric.

Ghosts

Picture Courtesy of General Electric.

Weak Signal Snow

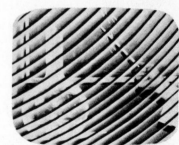

Picture Courtesy of General Electric.

Radio-Frequency Interference

Occasionally television reception will be unsatisfactory through no fault of your operation. Shown here are "ghosts," "snow," and R-F (radio-frequency) interference. Ghosts are usually caused by station signals reflected from obstacles between the transmitter and your antenna; snow is usually caused by weak signals, ofen the result of an inadequate or faulty antenna system; R-F interference may be caused by extraneous electrical signals. When these problems occur, be sure your antenna cables are sound and properly connected, and any antenna switches are in their proper positions. If these are all in order, call for technical assistance.

EXERCISE 65

VIDEO PLAYERS AND RECORDERS: DISC, CASSETTE, OPEN REEL

DoAll Company.

The Reason for It

A wide variety of prerecorded visual materials is available on video tape, video cassette, and video disc. Video tape's ability to record opens the possibility of storing information for later use. Successful use of these materials requires the ability to connect and operate the various machines needed for playback and record.

Before You Start

Brown, Lewis, and Harcleroad: *AV Instruction*, Chap. 11; Reference Section 1. This manual: Exercise 64; Correlated References, Section 41.

Purposes

1 To help you to learn to use various types of video players for the presentation of prerecorded materials

2 To help you learn to make and play back simple video recordings

Required Equipment and Materials

1 A television receiver/monitor.

2 Video player/recorders as follows:

  a Video cassette player or recorder

  b Open-reel video player or recorder

  c Video disc player

3 Prerecorded video instructional materials:

  a Videocassette

  b Open-reel

  c Video disc

4 A blank reel or cassette of video tape for one of the recording machines.

5 Automatic locating controllers for the videocassette and video disc machines.

6 Operating manuals for each piece of equipment, especially the automatic locating controllers.

Assignments

Have at least one other person work with you on the following assignments which require manipulation of equipment. Take turns doing each operation:

1 For each of the three types of video player (cassette, open reel and disc) use the operator's manual and the worksheet at the end of this exercise. Regardless of their configurations, all video player/recorders have certain features in common. Examine your machine(s) and the generalized machines pictured on the worksheet. On your machine(s), locate each feature listed on the worksheet; then give a brief description of its function in the space provided. Note items not present on your machine. Also note those items present on your machine but not listed on the worksheet.

2 After reading the remainder of this exercise, connect in turn each of the three types of machines to the TV monitor/receiver and play back a short segment of prerecorded material. First, connect the video and audio outputs of the machine to the video and audio inputs of the set. Then connect the R-F output of the machine to the antenna input of the set. What differences can you notice in the quality of picture and sound between the two modes of connection? Rotate the playback controls of each machine, such as skew, tracking, and color lock, through their range to become familiar with their effects on the picture.

3 Using the automatic locating devices for the cassette and disc machines, practice selecting a short segment of a prerecorded program, noting its address, removing the program from the machine, reinserting it, and locating selected segments with the controller. If your controller permits programming a number of segments to play in a specified order, practice this mode of operation.

4 Using one of the video tape or cassette units which has record capability, make a short recording of a broadcast TV program. (NOTE: Copyright laws restrict recordings of this type. What are your school district policies on off-the-air recordings?)

5 If your recorder is equipped with an "edit" button, read the section on editing in the operator's manual and then edit together a three- or four-scene sequence. Experiment by making edits with the tracking control set correctly and with it misadjusted. Note the importance of proper tracking adjustment when editing. Try various edits with roll times of 2 to 10 seconds. Note also the increased stability of an edit made with the longer roll times.

Video Player Operation

Video systems differ in their complexity and in their connections and controls. Operational steps described below are typical of most systems, whether open-reel, disc, or cassette. Remember that in many respects video machines operate similarly to audio machines; refer to your experience with audio tape recording as you learn to use the video tape recorder. Follow these procedures:

1 Set up and interconnect the player and the TV receiver. Either video/audio or R-F interconnect may be used for playback.

2 Plug in power cords and turn on the player and monitor.

3 For cassette and disc systems: open the loading mechanism and insert the program. For open-reel machines, make certain that the video heads are not turning, and thread tape onto the machine following directions in the operator's manual. Thread tape carefully; avoid creasing it or touching the recording surface.

4 Set the index counter to zero.

5 Play a portion of the program. The television receiver can now be adjusted in the manner shown in Exercise 64.

6 If your machine has controls for tracking and tension, adjust them. The tension control adjusts the top of the picture so that any vertical objects will be straight up and down and not bent to the left or right. Tracking is adjusted for the most steady, snow-free picture. Refer to your operating manual for details.

7 Return the tape or disc to the head of the program; it is now ready to play.

Handling and Storing Video Tapes and Video Cassettes

1 Avoid extremes of temperature or humidity such as a closed car trunk in summer or a damp basement in winter. Tape survives well at comfortable "human" temperatures.

2 Tape should be kept dust-free. Any dust or dirt on the tape surface will cause black or white flashes on the television screen. Use tape in a clean environment, and avoid smoking in video tape recording areas.

3 Return the tape to its box immediately after taking it off the recorder.

4 Store boxes vertically on shelves. Do not store tapes flat for any length of time.

5 Be sure to label your video tapes. Label both the reel and the box. Minimum label information should include: (a) the name of owner, (b) name of program, (c) type of video tape recorder or format used, (d) color or black and white, (e) original recording or a copy, and (f) length of program.

6 When using open reel video tapes, do not remove the tape from guides, capstan, or head drum when you have played partially through a program. To avoid serious damage to the tape, wind all tape onto either the supply or the take-up reel.

Handle open reel tape by reaching under the reel from two sides and lifting. Do not lift the top flange or squeeze flanges together. Cut off and discard ends of tape that have been creased or that are contaminated from other sticky tapes.

Video Recorder Operation

For video tapes and cassettes, recording is a simple extension of player operation:

1 Connect the system as if for playback plus:

a To record "Off-Air" from a television receiver/monitor, audio and video outputs from the receiver must be connected to the audio and video inputs of the recorder. Some videocassette recorders have built-in tuners which do not require this connection; in such cases the recorder's input selector switch should be set to its television position. A suitable television antenna must be connected to the television set or to the VTR tuner.

b If the source is to be a microphone and camera, they must be connected to the audio and video inputs on the recorder and the input selector must be set to the "line," or camera position.

2 Load tape onto the machine. With cassettes, it is necessary to check that the safety button or tab is in place on the cassette; otherwise, the machine will not go into record.

3 If audio and video level controls are provided, adjust them to obtain the recommended meter reading. Many cassette re-

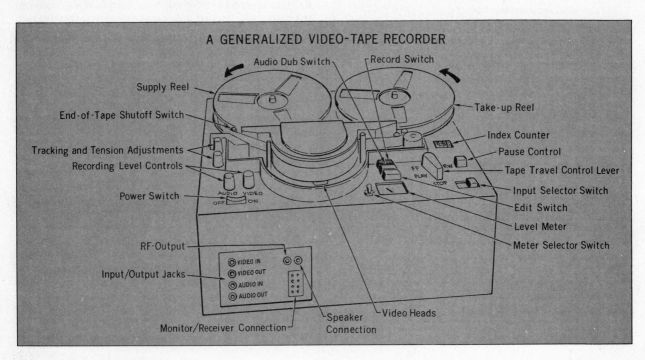

A GENERALIZED VIDEO-TAPE RECORDER

corders allow this adjustment to be made with the machine in the stop mode. Many reel-to-reel recorders require you to press the record button (without putting the tape in motion) in order to adjust these controls.

4 Make and play back a short test recording immediately before doing a complete recording to ensure that all components are operating properly. If your machine permits, go into the record mode directly from the forward (play) mode without first stopping the machine. Several advantages are gained by doing this. Since the machine is successfully playing back your test recording, you know that the tape is in good condition and is threaded properly, that the video and audio heads are clean, and that the motor drive controls are operating properly. With these conditions checked, by the simple procedure of going directly into the record mode from the play mode, you can eliminate most of the possible sources of

trouble before you start recording.

If your machine does not permit going into the record mode from playback without stopping, it may be possible to go into record from a pause mode or a stop-motion position. If the machine is satisfactorily reproducing a still frame of your test recording just before recording, you can, as above, have confidence your recording will in all probability be satisfactory.

When working with even simple video recording systems, difficulties inevitably will arise. As evidence of "gremlins" is encountered, it is wise to list problems on a trouble sheet as a guide to others who might encounter them also. Always, in consideration for the next operator, either arrange for faulty equipment to be repaired before it is put away or make a written report to ensure repair before the next scheduled use.

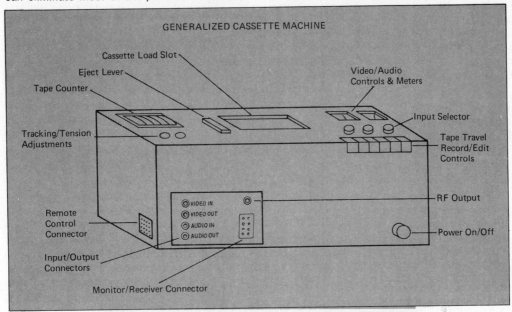

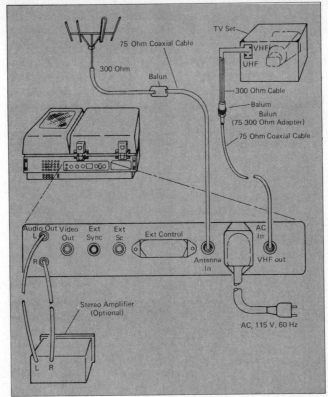

MCA Discovision.
Standard television set system.

Pictured above are three video formats: (1) U-Matic Video Tape (reel-to-reel); (2) generalized cassette, and (3) video disc.

WORKSHEET 65

**VIDEO RECORDER/PLAYBACK
EQUIPMENT CHECKLIST**

Name _____

Course _____ Date _____

YOU MAY COPY THIS FORM TO COMPLETE THE EXERCISE

Feature to Compare	Functions (Machines 1, 2, and 3)		
	1	2	3
Power switch			
Audio level control			
Video level control			
Level meter			
Meter selector switch			
Tape travel controls			
Record switch			
Insert switch			
Assemble switch			
Video input			
Audio input(s)			
Video output			
Audio output(s)			
Speaker connection			
Audio dub switch			
Monitor/receiver connection			
End-of-tape shutoff switch			
Edit switch			
Index counter			
Tension adjustment			
Track adjustment			
R-F output			
Eject lever			
Input selector			
Other			

152

TELEVISION CAMERAS

The Reason for It

Teachers in many fields will have occasion to use the close-up, image-magnification functions of TV systems to make details of demonstrations visible to large groups in classrooms, shops, laboratories, and lecture halls. When a camera is added to a video recorder you have a system which you can use to store information for later single or multiple playbacks and with which you may record and instantly replay student performances.

Before You Start

Brown, Lewis, and Harcleroad: *AV Instruction*, Chap. 12. This manual, Exercises 64 and 65; Correlated References, Section 41.

Purposes

1 To learn to use a single-camera television system for image magnification of demonstrations

2 To learn to make and play video tape recordings with a one-camera, portable video tape recorder system

Required Equipment and Materials

1 A portable television system, including: camera, zoom lens, close-up adaptor, several fixed-focal-length lenses, extension tubes, tripod and/or overhead support stand, and a TV receiver

2 All items in 1, and a portable video tape recorder, all necessary connecting cables, a microphone, and auxiliary lights, if required

3 A blank cassette or reel of video tape

4 Operating manuals for each piece of equipment, especially the video tape recorder

Assignments

Have at least one other person work with you on the assignments below, which require manipulation of equipment. Take turns doing each of the following operations:

Lee County School District, Florida; Jim Garrison.

1 Using the step-by-step procedure on the next page, set up a single-camera image-magnification system, and present a demonstration. Select flat materials, three-dimensional objects, and pages from books. How does your focus change in relation to the thickness of objects? Is a full page of print from a book legible on the screen, or do you have to show a closer view, a portion at a time? Working from different sides of the camera, try moving a flat object under the camera and pointing to its features while watching the monitor. Where do you have to be in relation to the camera bottom to make this manipulation easiest?

2 With a close-up adaptor on your zoom lens and/or extension tubes on a fixed lens, experiment with tight close-ups of both flat and three-dimensional objects. Is focus critical with tight close-ups? What problems do you have with three-dimensional objects? Is image steadiness a problem?

3 Using the experience gained in Exercise 65, video player/recorders, and the operator's manual for your recorder, make a recording using the VTR, camera, and microphone. During the recording, vary through their range the adjustments on the camera and recorder including: camera focus, lens focus, lens aperture, VTR audio level, and VTR video level. Into the microphone describe your actions so that they will be recorded on the tape to help you analyze the effect of your adjustments when you play the tape back.

Single-Camera System Operation

Although each single-camera system has features peculiar to it, all such systems have principles of operation in common. The procedure below outlines steps for operation of typical equipment. (Skills you may have in operating still and motion picture cameras will help you in using the television camera.)

1 Set up the camera on a tripod or stand. The camera can be supported on a tripod for close-up work. However, mounting on special stands, such as the one shown, greatly facilitates demonstration with this system, especially when an overhead position is desirable for a subjective, how-to-do-it, over-the-shoulder view.

2 Auxiliary lights may or may not be required to obtain a satisfactory picture. (Pointing the camera at any bright light may damage the costly camera tube.) Lighting will vary with subject matter. Arrangements such as those recommended for photographic copying are often suitable. If the subjects will be manipulated or pointed to, lights should be arranged so that any shadows from hands or pointers will fall away from, rather than on, the subject.

3 Connect the camera to the receiver with the cable and connectors required by the units; adjust the receiver to the R-F or video channel transmitted by the camera.

4 Turn on the camera and receiver units. If the system is not transistorized, allow several minutes for warmup.

5 Place an object or picture in the field of view of the camera lens.

6 Adjust the focus ring on the camera lens for the sharpest picture. If there is a focus adjustment at the rear of the camera, adjust this also. If you use a zoom lens, adjust it for the widest possible picture, and focus with the knob at the rear of the camera. Then zoom in to the tightest possible close-up, and adjust the focus ring on the lens. The picture should

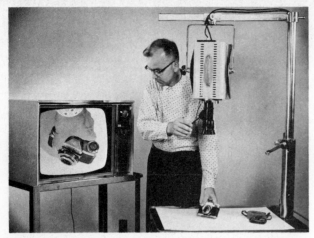

now stay in focus throughout the zoom range from wide shot to close-up for that subject. Note that when you shift the camera to a subject at a different distance, it will be necessary to refocus the lens focus ring.

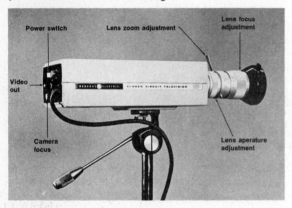

7 To adjust the lens for the best picture contrast, first adjust the receiver/monitor for a normal broadcast picture; then switch to the camera and adjust the lens aperture to produce a picture of quality similar to that of the broadcast picture. Note how the depth of field (range of distances from the camera that are in sharp focus) change when you change the aperture settings and when you use different lenses.

8 If yours is a color camera, and color balance adjustments are provided, place a white card before the camera so that it fills the screen and adjust the color balance controls according to your operating manual. The light falling on the white card should be the same light that will fall on the subject. If there is a change in color of the light on your scene, such as a change from incandescent indoor light to daylight, this adjustment must be performed again.

9 Make final adjustments on the receiver/monitor for brightness and contrast.

10 While watching the monitor, rehearse your presentation; practice arranging materials and using camera and lens adjustments; particularly lens aperture, zoom, and focus.

As with still cameras, tight close-ups with television cameras can be facilitated with close-up attachments and with extension tubes. These accessories allow the camera to be focused on objects much closer than the lens used alone would permit. (See Exercise 44.) Experiment with these devices in viewing three-dimensional as well as flat objects; evaluate the limits of detail that can be resolved on the monitor screen.

Recording from TV Cameras

As you learned in Exercise 65, recording systems differ in their complexity, connections, and controls. The steps described below are typical for most systems. In many respects television cameras operate similarly to motion picture and still cameras. Your experience with those will help you as you learn to use the TV camera. Follow these steps:

1 Set up and interconnect the system components (camera, video tape recorder, microphone, and monitor/receiver). To speed setup and minimize confusion in interconnecting system components, you may find it advantageous to color-code or label cables and their connecting points.

2 Plug in power cords and turn on the camera, recorder and monitor receiver.

3 As in Exercise 65, load the cassette or thread the tape following directions in the operator's manual. Set the recorder index counter to zero.

4 Remove the lens cap from the camera and aim the camera at the subject. Some recorders allow you to see the picture from the camera when the recorder is stopped. If yours does not, press the record button without putting the tape into motion. This will permit you to see the picture produced by the camera so that you can aim and adjust the camera before you start recording. Focus the camera lens and adjust the aperture, as described previously in this exercise. If video level control is provided on the recorder, adjust it to obtain the recommended meter reading.

5 If audio volume control is provided on the recorder, have someone speak into the microphone, adjust volume for a proper meter reading.

6 Put the machine into record mode (usually *forward* and *record*), and make a short test recording. Rewind and replay the tape. Observe the picture and sound. Readjust the record level controls if necessary.

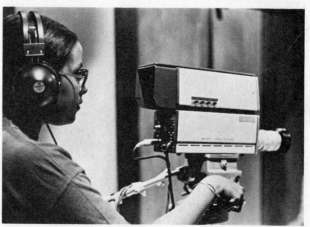

Virginia State College

EXERCISE 67

MICROCOMPUTER EQUIPMENT

Foothill College, California.

The Reason for It

The microcomputer is a general-purpose, interactive learning device. General purpose means that the computer will accommodate any content that the user wishes to choose, whether it be mathematics, social studies, reading, games, or science. Interactive means that the microcomputer must have commands from the user before it will do anything. The microcomputer also has the capacity to store, organize, and present great amounts of information in a very short time—often the fraction of a second.

Before You Start

Brown, Lewis, and Harcleroad: *AV Instruction*, Chap. 15. This manual: Exercise 13; Correlated References, Section 13.

Purposes

1   To learn the advantages and limitations of various kinds of microcomputers

2   To develop skill in start-up operations for microcomputers

Required Equipment and Materials

1   Several microcomputers of different brands using cassettes and disc drives with black and white and color monitors. (If your school does not have microcomputers, stores will generally let you practice on various types of machines.)

2   Cassette tape recordings and discs containing computer programs that may be run on the machines

3   Manuals that describe the operation of the machines you wish to examine or evaluate

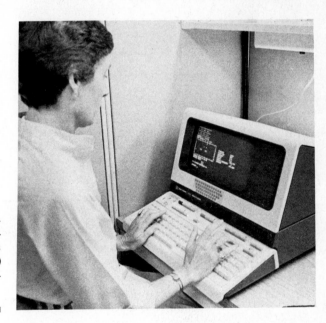

Assignments

1   Probably not more than a half-dozen manufacturers represent the major microcomputer market. Often, their representatives (dealers) will be willing to make demonstrations in local schools or computer stores. Study the operating manual's start-up instructions; try the machines with programs in your special areas of interest. Prepare an evaluation of each machine tested, using these criteria: ease of use, clear instructions in the manual, and quality of programs for classroom use. (See checklist on the following page.)

2   Check the contents of several publications (see the Correlated References section) for information about periodicals that deal with the microcomputer field. *Infoworld* and *Interface Age* are excellent prospects for this activity.

3   Visit schools in your area that make frequent (and functional) use of computers in the classroom. Discuss with teachers and students at each of these schools the pros and cons of this activity. Why did they turn to microcomputers? Have microcomputers met their expectations?

WORKSHEET 67

Name _____

Course _____ Date _____

YOU MAY COPY THIS FORM TO COMPLETE THE EXERCISE

Computer Brand Names

Unit Details							
TYPE OF UNIT	Computer and components in a single unit						
	Computer and separate components						
RAM—MEMORY SIZE:	4K						
	8K						
	16K						
	32K						
	48K						
	More						
INPUT/METHOD OF LOADING:	Cassette load						
	Single disc drive						
	Double disc drive						
	Other						
EXPANSION CAPABILITY:	Memory expandable						
	Memory not expandable						
SIZE OF SOFTWARE; LIBRARY AVAILABLE:	100 programs						
	500 programs						
	1000 programs						
	2500 programs						
	More						
TYPE OF SERVICE AVAILABLE:	Locally serviced						
	Serviced at factory						
	Remote-area serviced						
OUTPUT/DISPLAY	Black and white						
	Color						
	Printer						
LANGUAGES AVAILABLE:	BASIC						
	PASCAL						
	PILOT						
	FORTRAN						
	COBOL						
	CP/M						
	Other						
WARRANTY:	30-day warranty						
	90-day warranty						
	1-year warranty						
	Extended warranty						
COST:	Basic system						
	With peripherals						
	Memory size						

Rate the microcomputers you have evaluated in order of preference for: (1) Ease of operation, (2) quality of manual, (3) size and quality of program library, and (4) cost.

EXERCISE 68

EQUIPMENT MAINTENANCE

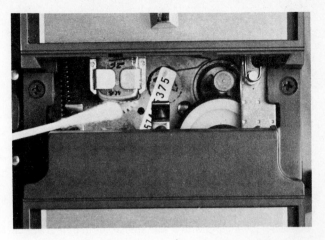

The Reason for It

Teachers usually are not expected either to maintain audiovisual equipment or to perform major repairs on it. However, there are numerous routine emergencies that competent operators, including teachers, must meet and must solve. Operators ought to be able to diagnose simple equipment failures and distinguish such emergencies from operator errors. This exercise includes not only pointers on maintenance but also items in equipment operation that help eliminate calls for assistance.

Before You Start

Brown, Lewis, and Harcleroad: *AV Instruction*, Reference Section 1. This manual: Correlated References, Section 16.

Purposes

1 To learn to diagnose equipment failures

2 To learn to avoid operator errors often interpreted as equipment failures

3 To learn minor maintenance procedures for audiovisual equipment

4 To learn to replace projection and exciter lamps

5 To understand equipment grounding systems

Required Equipment

1 Motion picture equipment and accessories

2 Filmstrip projector

3 Tape recorder

4 Record player

5 VTR

6 Television receiver/monitor

Required Materials

1 Motion picture film

2 Spare lamps for projectors

3 Lens tissues

4 Alcohol (denatured)

5 Cotton-tipped swabs

6 Audio and video tape

Assignments

Study the tips on equipment care and maintenance in this exercise; then perform these tasks:

1 *Motion picture projection equipment:*

a Remove and replace the projection and exciter lamps.

b Remove, inspect, and clean, if necessary, a projector lens.

c Have your associates disable a projector by misadjustment, misconnection, or other nondamaging treatment. Determine how quickly you can locate the trouble.

d Examine the entire film path through the projector. Determine all points where regular cleaning will protect the film.

e If the projector is equipped with steel spring belts that drive the reels, remove and replace one of them.

2 *Still-projection equipment:*

a Remove and clean the projection lens. Test for proper replacement.

b Remove and replace the projection lamp.

3 *Audio equipment:*

a Inspect and determine the condition of all cords.

b Inspect the stylus (needle) of a record player; be sure all wires are attached to it.

c Clean the heads of a tape recorder. Replace all covers properly.

4 *Television equipment:*

a As with the motion picture projector assignment, above, have associates disable both the video tape recorder and television monitor by misadjustment of readily accessible external controls and/or cable disconnections. Determine how quickly you can locate the trouble or troubles.

5 *Any equipment:*

a Check the condition of connecting cords, and indicate where repair or preventive maintenance is needed.

b Check for loose knobs, screws, handles, and accessories. Determine how these should be tightened or adjusted.

Projector Lamp Replacement

1 Be sure the lamp is cool before attempting to change it. Use a clean cloth, tissue, or white glove to avoid leaving fingerprints on the lamp surface, which would, in turn, lead to early failure.

2 Always replace a lamp with another of the proper specified base, filament, glass envelope shape, wattage, and voltage.

3 Never use a lamp with a wattage larger than that specified (usually on a metal tag of the projector housing).

4 If a projector has no fan or blower, never use a lamp larger than 300 watts.

5 When changing lamps, first turn off the projector power and disconnect the power cord from the power source. Open lamp housing. Remove lamp. Notice that, unlike lamps in the home, the lamp does not unscrew. Projection bulbs have bases that absolutely control their positions so that their filaments distribute light evenly.

Lens Care

1 *Never* touch the glass elements with your fingers.

2 Clean lenses *only* when they are dirty.

3 To clean a lens, you may blow on it, or breathe on it, then wipe gently with *lens tissue* (and with nothing else).

4 Use a clean piece of lens tissue every time you clean a lens. (Silicone-type tissue for eyeglasses is not recommended for projector lenses.)

5 Leave cleaning of condensers and reflecting mirrors to your audiovisual service office.

6 Never attempt to take a lens apart.

Audio Equipment Care

Lack of sound may be caused by:

1 Power cord not plugged in

2 Amplifier not on

3 Speaker not connected

4 Any cord not fully plugged in or plugged in the wrong jack

5 Defective equipment, such as burned-out exciter lamp in motion picture projector

The above hints are only samples. But note that faulty operation is always a possibility, often a probability, *rather than faulty equipment*.

Cord plugs may fail because wires inside them may break, thus causing intermittent sound. This happens especially where the cord enters the plug or the equipment itself. If sound is intermittent, shake the cord and listen for tell-tale crackling and pronounced intermittent sounds.

Cleaning Audio Tape Recorders

1 In addition to keeping external portions of the tape recorder clean, it is imperative that the heads of the machine be clean.

2 Use a cotton-tipped swab and only a *little* alcohol. Rub lightly.

3 *Never* use metal to clean heads. High-quality performance depends upon this routine cleaning; a garbled or otherwise unsatisfactory recording *may* be caused by dirty heads.

4 Also use alcohol to clean other audio tape recorder mechanisms, such as the tape drive rollers. While other cleaners sometimes are recommended for this purpose, they may *damage* tape heads; therefore, avoid any solvent except those specified by the manufacturer.

Caring for Television Equipment

When television equipment does not operate properly, follow these steps:

1 Are all cables connected properly? Many cables can be plugged in at more than one place. Are your cables connected where they should be to do the job you want to do?

2 Check all switches and control knobs to make certain each is in the correct position.

3 Think through the chain of events necessary to accomplish the end you want: (a) light from the subject passes through the camera lens, (b) the image is transmitted from the camera to the recorder and onto the tape, then from the tape through the recorder again to the monitor. What could interrupt this path? Is the lens cap on? Is the lens iris closed? Is the lens or camera out of focus? Are the recorder video heads dirty? And so on.

4 If possible, substitute similar components, such as cables, camera, and recorder, to isolate the trouble to a particular component.

5 To help in troubleshooting, keep a tape on hand which has good-quality audio and video recording and is at least 5 minutes in length.

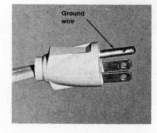

Ground Wire Systems

Some power plugs have a third wire called a ground wire. The purpose of the ground wire is to prevent electrical shock in the event of a short circuit occurring in the equipment. Grounding of electrical equipment in public schools is sometimes required by law. Know the law as it applies to you.

When the electrical outlets are designed for two-prong plugs and the plug has a third prong, an adapter is available to enable you to provide grounding protection. The wire with the horse-shoe connector is the ground wire. The screw that holds the cover plate of the outlet in place may be grounded. Check with media staff to find out. If it is grounded, attach the horse-shoe connector behind the screw.

The Audiovisual Equipment Service Center

Although in emergencies teachers may need to make simple adjustments or minor repairs to equipment, most school systems provide audiovisual equipment repair and maintenance services; other districts may contract with local audiovisual dealers for equipment servicing. Work on equipment requires special tools and test instruments, a supply of lamps and spare parts, and convenient bench space where competent technicians can do electrical, electronic, and mechanical operations. Become acquainted with service facilities available to you and report your needs for repairs to technicians there.

SECTION FOUR

CORRELATED REFERENCES

Included here are listings of various media—books, filmstrips, audio tapes, motion picture films, video tapes, records, transparency sets, multi-media kits, as well as others—that correlate with one or more exercises of this manual. They are arranged alphabetically by author (for print materials) or by title (for audiovisual items) under headings that occur in the following order:

1 Sources of current information
2 General references
3 Audio: Records
4 Audio: Tapes
5 Boards: Bulletin
6 Boards: Chalk
7 Boards: Cloth, Magnetic
8 Boards: Electric
9 Books: Reference
10 Books: Textbooks
11 Charts and Chartmaking
12 Community Resources
13 Computers
14 Displays
15 Duplicating: Equipment, Processes
16 Equipment: General
17 Equipment: Audio
18 Equipment: Motion Picture
19 Equipment: Overhead Transparency
20 Equipment: Slide and Filmstrip
21 Equipment: Video
22 Equipment: Special Types
23 Facilities
24 Filmstrips
25 Free, Inexpensive, Sponsored Materials
26 Games and Simulations
27 Instructional Design and Systems
28 Learning Centers, Media Centers, Resource Centers
29 Lettering
30 Maps and Globes
31 Modules and Kits
32 Motion Pictures: Film and Video
33 Photography: Motion
34 Photography: Still
35 Puppets and Puppetry
36 Selection Aids: General
37 Sketching
38 Slides: 2" x 2"
39 Still Pictures: Flat, Opaque
40 Still Pictures: Laminating, Mounting
41 Television
42 Transparencies: Large
43 Visual Literacy

Names and addresses of manufacturers and suppliers of audiovisual equipment, audiovisual supplies (for lettering, mounting, or producing transparencies, for example), and various media software such as films or filmstrips, textbooks, tape or disc recordings, or maps, globes, and charts have not been included here. To obtain this information, consult Reference Section 5, "Classified Directory of Sources," in *AV Instruction: Technology, Media, and Methods* (6th ed.), by James W. Brown, Richard B. Lewis, and Fred F. Harcleroad; McGraw-Hill, N.Y., 1983.

1 Sources of Current Information

Although you will find in this list of correlated references sources of data needed as background or as supplementary information for the various assignments given, you should also give attention to other materials that will be produced or published after this list goes to press. For this purpose we suggest a number of regularly updated items, including the following:

Periodicals

• *ECT Journal* (Association for Educational Communications and Technology, 1126 16th St. N.W., Washington, D.C. 20036).

• *Educational Technology* (Educational Technology Publications, Inc., Englewood Cliffs, N.J. 07632).

• *Electronic Learning* (902 Sylvan Ave., Box 2001, Englewood Cliffs, N.J.).

• *ERIC/IR Update* (ERIC Clearinghouse on Information Resources, School of Education, Syracuse University, Syracuse, N.Y. 13210).

• *Instructional Innovator* (Association for Educational Communications and Technology, 1126 16th St. N.W., Washington, D.C. 20036).

• *Library Journal* and *School Library Journal* (R. R. Bowker Co., 1180 Ave. of the Americas, New York 10036).

• *Media and Methods* (American Society of Educators, 1511 Walnut St., Philadelphia 19102).

• *School Library Media Quarterly* (American Association of School Librarians, 50 E. Huron St., Chicago 60611).

• *Sightlines* (Educational Film Library Association, 43. W. 61st St., New York 10023).

• *Training and Development Journal* (American Society for Training and Development, 600 Maryland Ave. SW, Washington, D.C. 20024).

General Sources

- *The Audio-Visual Equipment Directory:* National Audio-Visual Association, Fairfax, Va., revised annually.

- *Current Index to Journals in Education.* Phoenix, Ariz.: Oryx Press, mo., 2 semi-annual cumulations.

- *Resources in Education.* Phoenix, Ariz.: Oryx Press, 3 vols. plus annual supplement.

- *Education Index.* New York, NY: H.W. Wilso Co., 10/yr.; cumulations.

- *Educational Media Yearbook,* James W. and Shirley N. Brown, eds. Littleton, Col.: Libraries Unlimited, Inc., ann. The 1982 edition carries addresses of most of the publishers and a large number of the producers in the educational media field. Future editions will provide updates of new materials related to the topics included in this "Correlated References" section.

2   General References

The following general textbooks were available for use in introductory audiovisual or instructional technology classes at the time this manual was published:

Brown, James W., Richard B. Lewis, and Fred F. Harcleroad: *AV Instruction: Technology, Media, and Methods* 6th ed. McGraw-Hill, New York, 1983.

Bullough, Robert V.: *Creating Instructional Materials* (2d ed.), Charles E. Merrill Publishing Co., Columbus, Ohio, 1978.

Gerlach, Vernon S., and Donald P. Ely: *Teaching and Media: A Systematic Approach* (2d ed.) Prentice-Hall, New York, 1980.

Haney, John B., and Eldon J. Ullmer: *Educational Communications and Technology: An Introduction* 3d ed. Wm. C. Brown Publishers, Dubuque, Iowa, 1980.

Heinich, Robert, Michael Molenda, and James D. Russell: *Instructional Media and the New Technologies of Instruction.* John Wiley and Sons, New York, 1982.

Sleeman, Phillip J., Ted S. Cobun, and D. M. Rockwell: *Instructional Media and Technology: A Guide to Accountable Learning Systems.* Longman, Inc., New York, 1978.

Wittich, Walter A. and Charles F. Schuller: *Instructional Technology: Its Nature and Use* 6th ed. Harper and Row, New York, 1979.

3   Audio: Records

Adventures in Observing and Listening, 35mm filmstrip with cassette, January Productions, Hawthorne, N.J., 1980.

Index to Educational Records, National Information Center for Educational Media, Los Angeles, regularly revised.

Recordings for Children, New York Library Association, New York, 1981.

Sound Recording and Reproduction, Focal Press, New York, 1981

Schwan Record and Tape Guide, Schwan, New York, regularly revised.

4   Audio: Tapes

Audiofunics, kit. Prentice-Hall Learning Systems, San Jose, 1980.

Audio Cassette Directory. Cassette Information Services, Glendale, Calif., 1982.

Index to Audio Tapes. National Information Center for Educational Media, Los Angeles, revised biennially.

Catalog. National Center for Audio Tapes, Bureau of Audio-visual Instruction, University of Colorado, Boulder.

Complete Handbook of Magnetic Recording, TAB Books, Inc., Blue Ridge Summit, Pa., 1981.

Guide to Making an Audio Tape, microfiche, ERIC Document Reproduction Service. (ED 171 314). ERIC, Washington, D.C., 1978.

Schwan Record and Tape Guide, New York: Schwan, revised regularly.

5   Boards: Bulletin

Bulletin Board Ideas, Kentucky State Department of Education, Frankfort, 1978.

Bullough, Robert V.: *Display Boards,* Educational Technology Publications, Inc., Englewood Cliffs, N.J., 1981.

Coplan, Kate: *Poster Ideas and Bulletin Board Techniques,* Oceana Publications, Inc., Dobbs Ferry, N.Y., 1981.

Franklin, Linda C.: *Library Display Ideas,* McFarland and Co., Jefferson, N.C., 1980.

Gill, Bob: *Forget All the Rules You Have Ever Learned about Graphic Design,* Watson-Guptill, New York, 1980.

Posters, Children's Books Council, New York, 1981.

6   Boards: Chalk

Gordon, Robert M., "Classrooms Without Chalkboards," *Educational Technology,* April 1975, pp. 38-40.

Improving the Use of the Chalkboard, 35mm filmstrip, color. Columbus, Ohio: Bureau of Educational Research, 1970.

7   Boards: Cloth, Magnetic

Basic Educational Graphics (Multimedia Series): *Flannel Board Construction* and *Flannel Board Use,* 35mm filmstrips, Scott Educational Division, Holyoke, Mass., 1970.

Frye, Roy: *Graphic Tools for Teachers,* Graphic Tools for Teachers, Mapleville, R.I., 1975.

Donahue, Bud: *Language of Layout,* Prentice-Hall, New York, 1978.

8   Boards: Electric

Morlan, John E.: *Preparation of Inexpensive Teaching Materials.* Crowell, New York, 1973.

Morlan, John, and Leonard Espinosa: *Electric Boards You Can Make.* Personalized Learning Associates, San Jose, 1974.

9   Books: Reference

Deveny, Mary Alice: *Recommended Reference Books in Paperback,* Libraries Unlimited, Inc., Littleton, Col., 1981.

Walford, A.J.: *Guide to Reference Materials,* American Library Association, Chicago, Ill., 1980.

Wynar, Bohdan S. (ed.): *American Reference Books Annual,* Libraries Unlimited, Inc., Littleton, Col.

Wynar, Bohdan S.: *Recommended Reference Books for Small- and Medium-Sized Libraries and Media Centers,* Libraries Unlimited, Inc., Littleton, Col., 1981.

Wynar, Christine: *Guide to Reference Books for School Media Centers,* Libraries Unlimited, Inc., Littleton, Col., 1981.

Wynar, Christine: *Index to American Reference Books Annual,* Libraries Unlimited, Inc., Littleton, Col., 1980.

10   Books: Text

Book Publishing and Distribution, Practicing Law Institute, New York, 1979.

Brown, Lucy Gregor: *Core Media Collection for Elementary Schools and Core Media Collection for Secondary Schools,* R. R. Bowker Co., New York, 1981.

Eisenstein, Alizabeth: *Printing Press as an Agent of Change,* Cambridge University Press, Cambridge, Mass., 1979.

Goodbye, Gutenberg, 16mm sound film, 20 min. WNET Thirteen Media Services, New York, 1981.

Rice, Stanley: *Book Design: Systematic Aspects,* R. R. Bowker Co., New York, 1978.

Rice, Stanley: *Book Design: Text Format Models,* R. R. Bowker Co., New York, 1978.

Story of a Book, 16mm film, 16 min., Pied Piper, Inc., Verdugo City, Calif., 1980.

Winckler, Paul A.: *History of Books and Printing,* Gales Research Co., Detroit, 1979.

Woodbury, Marda: *Selecting Materials for Instruction: Issues and Policies; Selecting Materials for Instruction: Media and the Curriculum;* and *Selecting Materials for Instruction: Subject Areas and Implementation,* Libraries Unlimited, Inc., Littleton, Col., 1979, 1980.

World of Print, 16mm film, 17 min., International Film Bureau, Chicago, Ill., 1979.

11 Charts and Chartmaking

Brinton, Willard C.: *Graphic Methods for Presenting Facts,* Engineering Magazine Co., New York, 1914.

Dayton, Deane K.: "Computer-Assisted Graphics," *Instructional Innovator,* Sept. 1981, pp. 16–18.

Lockwood, Arthur: *Diagrams: A Visual Survey of Graphs, Maps, Charts, and Diagrams for the Graphic Designer,* Watson-Guptill, New York, 1969.

Merrill, Paul F., and C. Victor Bunderson, "Preliminary Guidelines for Employing Graphics in Instruction," *Journal of Instructional Development,* Summer 1981, pp. 2–9.

Minor, Ed: *Handbook for Preparing Visual Media,* McGraw-Hill, New York, 1978.

Satterthwait, Les: *Graphics: Skills, Media, and Materials,* Kendall-Hunt Co., Dubuque, 1977.

12 Community Resources

Recorded Messages as a Way to Link Teachers and Parents, Evaluation Report, CEMREL, Inc., St. Louis, Mo., 1981.

Nigg, Heinz, and Graham Wade: *Community Media,* Centre for Advanced TV Studies, London, 1980.

Sandak, Cass R.: *Museums: What They Are and How They Work,* Franklin Watts, Inc., New York, 1981.

Smith, Lynn: *This Is Your Museum Speaking,* 16mm sound film, 13 min. National Film Board of Canada, Ottawa, 1981.

Wood, Ruon Kent: *Community Resources,* Educational Technology Publications, Inc., Englewood Cliffs, N.J., 1980.

Yellow Pages of Learning Resources, MIT Press, Cambridge, Mass., 1976.

Zelmer, A. C. Lynn: *Community Media Handbook,* Scarecrow Press, Metuchen, N.J., 1978.

13 Computers

Advanced Basic Techniques, a kit, Educational Activities, Inc., Freeport, N.Y., 1981.

At Home with the Computer: What Can It Do for You?, video tape, Center for the Humanities, Communications Park, N.Y., 1981.

Basic Computer Literacy (series), sound filmstrips. Eye Gate Media, Jamaica, N.Y., 1981.

Bass, George M.: *Creativity Through the Microcomputer* (ERIC ED 190 128), microfiche, ERIC, Washington, D.C., 1980.

Chickering, Arthur W.: *Experience and Learning: An Introduction to Experiential Learning,* Change Magazine Press, New Rochelle, 1981.

Douglas, Shirley, and Gary Neights: *Microcomputer References: A Guide to Microcomputers* (ERIC ED 205 203), ERIC, Washington, D.C., 1981.

Feldman, Phil, and Tom Rugg: *32 Basic Programs for the Apple Computer.* dilithium Press, Beaverton, Oregon, 1981.

Frenzel, Louis E. Jr.: *Howard W. Sams' Crash Course in Microcomputers,* Data Dynamics Technology, Cerritos, Calif., 1981.

Gosling, P. F.: *Beginning BASIC,* Robiotics Press, Forest Grove, Oregon, 1980.

Jenkinson, Edward B.: *Censors in the Classroom: The Mind Benders,* Southern Illinois Univ. Press, Carbondale, 1980.

Kellisch, Frederick J.: "Computer Graphics on a Shoestring," *Instructional Innovator,* Sept. 1981, pp. 19–23.

Marsch, Paul (ed.): *Guidelines for Post-Secondary Learning Resources Programs.* Association for Educational Communications and Technology, Washington, D.C.: 1980.

Melmed, Arthur S.: "Information Technology for U.S. Schools," *Phi Beta Kappan,* January 1982, pp. 308–311.

Watson, William Scott: *67 Ready-to-Run Programs in BASIC.* TAB Books, Inc., Blue Ridge Summit, Pa., 1981.

Willis, Jerry, and Merl Miller: *Computers for Everybody.* dilithium Press, Beaverton, Oregon, 1981.

14 Displays

Bruno, Sal: *Art Elements: An Introduction,* 35mm sound filmstrip, BFA Educational Media, Santa Monica, Calif., 1981.

Drawing People: An Introduction to Figure Drawing, 35mm sound filmstrips, Educational Dimensions Group, Freeport, N.Y., 1981.

Poster. Clearinghouse on Development Communication, Washington, D.C., 1981.

15 Duplicating: Equipment, Processes

Johnson, Mark: *Film/Video Education and the New Copyright Act* (ERIC ED 203 846), ERIC, Washington, D.C., 1980.

Lawrence, John S., and Bernard Timberg (eds.): *Fair Use and Free Inquiry: Copyright Law and the New Media,* ABLEX Publishing Corp., Norwood, N.J., 1980.

Liquid Duplicating Systems, Wm. C. Brown Publishers, Dubuque, 1975.

Reprographics, set of large transparencies, Lansford Publishing Co., San Jose, 1975.

Stencil Duplicating Systems, Wm. C. Brown Publishers, Dubuque, 1975.

16 Equipment: General

EPIE, Educational Products Information Exchange, Box 620, Stony Brook, N.Y. 11791, publishes technical test reports for various types of audiovisual equipment.

Library Technology Reports. Chicago: American Library Association, regularly published.

National Audio-Visual Association (NAVA), 3150 Spring St.,

Fairfax, Va. 22031, publishes, annually, the *Audio-Visual Equipment Directory* containing illustrations and data about most items of audiovisual equipment; excellent source list.

17 Equipment: Audio

Alkin, Glen: *Sound Recording and Reproduction.* Focal Press, New York, 1981.

Alten, Stanley R.: *Audio in Media*, Wadsworth Publishing, Belmont, Calif., 1981.

Jorgensen, Finn: *Complete Handbook of Magnetic Recording*, TAB Books, Inc., Blue Ridge Summit, Pa., 1981.

Olson, Linda: "Technology Humanized: The Rate-Controlled Tape Recorder," *Media and Methods*, January 1979, p. 67.

18 Equipment: Motion Picture

Beatty, LaMond: *Motion Pictures*, Educational Technology Publications, Englewood Cliffs, N.J., 1980.

Bensinger, Charles: *The Home Video Handbook*, Video-Info Publications, Inc., Santa Fe, N.M., 1980.

Bensinger, Charles: *Video Guide*, Video-Info Publications, Santa Fe, N.M., 1981.

Hybett, Harry: *Complete Handbook of Videocassette Recorders*, TAB Books, Inc., Blue Ridge Summit, Pa., 1981.

Sigel, Efrem, et al.: *Video Discs: The Technology, the Applications, and the Future*, Van Nostrand Reinhold Co., New York, 1981.

Wilshusen, John: "How to Prevent Equipment Failures," *Instructional Innovator*, March 1980, pp. 35–36.

19 Equipment: Overhead Transparency

Basic Tips on Producing and Using Overhead Transparencies, National Audio Visual Association, Fairfax, Va., 1981.

I Like the Overhead Projector Because . . ., 35mm sound filmstrip, Association for Educational Communication and Technology, Washington, D.C., 1977.

Johnson, Roger: "Overhead Projectors: Basic Media for Community College," *Audiovisual Instruction*, March 1978, pp. 21–22.

Overhead Projection Guide, Arkwright Publications, Fiskerville, R.I., 1980.

Sparks, Jerry D.: *Overhead Projection*, Educational Technology Publications, Inc., Englewood Cliffs, N.J., 1981.

Use of the Overhead Projector and How to Make Do-It-Yourself Transparencies, Swan Pencil Co., North Fort Myers, Fla., 1981.

20 Equipment: Slide and Filmstrip

Beatty, LaMond F.: *Filmstrips*, Educational Technology Publications, Inc., Englewood Cliffs, N.J., 1979.

Beatty, LaMond F.: *Still Pictures*, Educational Technology Publications, Inc., Englewood Cliffs, N.J., 1981.

Kueter, Roger A.: *Slides*, Educational Technology Publications, Inc., Englewood Cliffs, N.J., 1981.

Saxby, Graham: *Focalguide to Slides*, Focal Press, New York, 1979.

21 Equipment: Video

Bensinger, Charles: *Video Guide*, Video-Info Publications, Santa Fe, N.M., 1981.

Sigel, Efrem, et al.: *Video Discs: The Technology, the Applica-*

tions, and the Future, Van Nostrand Reinhold, Inc., New York, 1981.

Harwood, Don: *Everything You Always Wanted to Know About Portable Videotape Recordings*, VTR Publishing, Syossett, N.Y., 1980.

Maltin, Leonard, and Allen Greenfield: *Complete Guide to Home Video*, Harmony Books, New York, 1981.

Robinson, Joseph, and Stephen Lowe: *Videotape Recording*, Focal Press, New York, 1981.

22 Equipment: Special Types

Bullard, John R., and Calvin E. Mether: *Audiovisual Fundamentals*, Wm. C. Brown Co., Dubuque, Iowa, 1979.

History of Microfilm, slide/cassette set, International Micrographic Congress, Bethesda, Md., 1980.

Kish, Joseph L. Jr.: *Micrographics: A User's Manual*, Wiley Interscience, New York, 1980.

23 Facilities

Cleaver, Betty: "A Media Facility Make-Over," *Instructional Innovator*, March 1981, pp. 20–23.

"Designing Facilities as If People Mattered," *Instructional Innovator*, March 1981, pp. 12–13.

Knirk, Frederick G.: *Designing Productive Learning Environments*, Educational Technology Publications, Inc., Englewood Cliffs, N.J., 1979.

Sullivan, Janet: "Nine Design Factors for Better Learning," *Instructional Innovator*, March 1981, pp. 14–16.

24 Filmstrips

Beatty, LaMond F.: *Filmstrips*, Educational Technology Publications, Inc., Englewood Cliffs, N.J., 1979.

Educators' Guide to Free Filmstrips, Educators' Progress Service, Randolph, Wisc., annual.

Index to 35mm Filmstrips (NICEM), National Information Center for Educational Media, Los Angeles, Calif., biennially.

25 Free, Inexpensive, Sponsored Materials

Educators' Guide to Free: (1) Audio and Video Materials; (2) Films; (3) Filmstrips; (4) Guidance Materials; (5) Health, Physical Education, and Recreation Materials; (6) Science Materials; (7) Social Science Materials; and *(8) Teaching Materials*, Educators' Progress Service, Randolph, Wisc., revised annually.

Elementary Teachers' Guide to Free Curriculum Materials, Educators' Progress Service, Randolph, Wisc., revised annually.

Lesko, Matthew: *Something for Nothing*, Associated Press Newsfeatures, Teaneck, N.J., 1981.

Vertical File Index, H.W. Wilson Co., New York, annual.

26 Games and Simulations

Blumenthal, Howard J.: *Complete Guide to Electronic Games*, New American Library, Bergenfield, N.J., 1981.

Gibbs, Emily A., and Jim Perry: *40 Computer Games*, Wayne Green, Inc., Peterborough, N.H., 1980.

Group Dynamics: Groupthink, 16mm sound film, 22 min., CRM/McGraw-Hill Films, Del Mar, Calif., 1980.

Jaffke, Freya: *Making Soft Toys*, Celestial Arts, Millbrae, Calif., 1981.

Lamoitier, Jean-Pierre: *Fifty Basic Exercises,* Sybex, Berkeley, Calif., 1981.

More New Games, New Games Foundation, San Francisco, Calif., 1981.

Puppet Program, kit with audio cassettes, Pretend Time Puppets, Inc., New York, 1981.

Puppets and You in the Library, Orange County Public Library, Orange, Calif., 1981.

Wilson, S. S.: *Puppets and People,* Tantivy Press, Ltd., London, 1981.

27 Instructional Design and Systems

Christiansen, James: *Criterion-Referenced Assessment: A Mini-Course* (ERIC ED 177 810), ERIC, Washington, D.C., 1979.

Eble, Kenneth E.: *Improving Teaching Styles,* Jossey-Bass, San Francisco, Calif., 1980.

Emmert, Phillip, and William C. Donaghy: *Human Communication,* Addison-Wesley, Reading, Mass., 1981.

Gagne, Robert M., and Leslie J. Briggs: *Principles of Instructional Design,* Holt, Rinehart, and Winston, New York, 1974.

Handbook of Procedures for the Design of Instruction, American Institutes for Research, Pittsburgh, Pa., 1978.

Kemp, Jerrold E., et al.: "Fads, Fallacies, and Failures," *Instructional Innovator,* January 1980, pp. 25–27.

Langdon, Danny G.: "Instructional Designs as Formats for Learning," *Instructional Innovator,* October 1981, pp. 24–26.

Orlich, Donald D., et al.: *Teaching Strategies: A Guide to Better Instruction,* D. C. Heath Co., Lexington, Mass., 1980.

Romiszowski, A. J.: *Designing Instructional Systems: Decision Making in Course Planning and Curriculum Design,* Nichols Publishing Co., New York, 1981.

Solomon, Gavriel: *Interaction of Media, Cognition, and Learning,* Jossey-Bass, San Francisco, 1979.

Sullivan, Janet: "Nine Design Factors for Better Learning," *Instructional Innovator,* March 1981, pp. 14–16.

Training in Business and Industry, Phi Delta Kappa, Bloomington, Ind., 1980.

Wildman, Terry M., and John K. Burton: "Integrating Learning Theory with Instructional Design," *Journal of Instructional Development,* Spring 1981, pp. 5–14.

28 Learning Centers, Resource Centers, Media Centers

Cabeceiras, James: *The Multi-Media Library: Materials Selection and Use,* Academic Press, New York, 1978.

Evaluating Media Programs: District and School, Association for Educational Communications and Technology, Washington, D.C., 1980.

Media Programs: District and School, Educational Communications and Technology, Washington, D.C. 1975.

Sheppard, George: *Using the School Media Center to Improve Instruction,* Idaho State University, Pocatallo, 1981.

29 Lettering

The following companies provide some printed materials pertaining to lettering equipment, materials, and techniques:

Artype, Inc., 345 E. Terra Cotta Ave., Crystal Lake, Ill. 60014.

Charles Meyer Studios, 168 Market St., Akron, Ohio 44308.

Kroy Industries, P.O. Box 43716, St. Paul, Minn., 55164.

Mittens Designer Letters, 345 5th St., Redlands, Calif. 92373.

Prestype, Inc., 194 Veterans Bldg., Carlstadt, N.J. 07072.

Reynolds/Leteron Co., 6704 Valjean Ave., Van Nuys, Calif. 91406.

Varigraph, Inc., P.O. Box 690, Madison, Wisc. 53791.

30 Maps and Globes

Inquire of the following producers of graphics (map and globe materials, especially) for booklets and other data sheets on utilization techniques.

George F. Cram Co., Inc., 301 South LaSalle St., Indianapolis, Ind. 46206.

Denoyer-Geppert Co., 5235 Ravenswood Ave., Chicago, Ill. 60640.

National Geographic Society, 17th and M Sts. N.W., Washington, D.C. 20036.

A. J. Nystrom Co., 3333 Elston Ave., Chicago, Ill. 60618.

Rand McNally and Co., 8255 Central Park Ave., Skokie, Ill. 60076.

31 Modules and Kits

Inquire of the following selected producers of kits and modular instructional materials for information regarding their offerings. Teachers' manuals are sometimes available.

Guidance Associates, Communication Park, Box 3000, Mt. Kisko, N.Y. 10549.

Interpretive Education, 400 Bryant St., Kalamazoo, Mich. 49001.

Jabberwocky, 4 Commercial Blvd., Suite 2, Novato, Calif. 94947.

Scholastic Book Services, 904 Sylvan Ave., Englewood Cliffs, N.J. 07632.

Science Research Associates, Inc., 155 N. Wacker Dr., Chicago, Ill. 60606.

Society for Visual Education, Inc., 1345 Diversey Parkway, Chicago, Ill. 60614.

32 Motion Pictures: Film and Video

The Catalog, Televised Higher Education, Boulder, Col. revised annually.

Index to Educational Films; Index to Educational Video Tapes; and *Index to 8mm Motion Cartridges,* National Information Center for Educational Media (NICEM), Los Angeles, revised biennially.

Educational Film Locater, Consortium of University Film Centers, R. R. Bowker Company, New York, revised annually.

Matoian, John: *Teaching Basic Skills with Film,* videotape master, Marlin Motion Pictures, Ltd., Mississouga, Ontario, 1981.

Mercer, John: *The Information Film,* Stipes Publishing Co., Champaign, Ill., 1981.

Short, Kenneth L.: *Feature Films as History,* University of Tennessee Press, Knoxville, Tenn., 1981.

Sigel, Efrem, et al.: *Video Discs: The Technology, the Applications, and the Future,* Van Nostrand Reinhold Co., New York, 1981.

33 Photography: Motion

Bullough, Robert V.: *Photography,* Educational Technology Publications, Englewood Cliffs, N.J., 1981.

Kemp, Jerrold: *Planning and Producing Audiovisual Materials,* Harper and Row, New York, 1980.

More Joy of Photography, Eastman Kodak Co., Rochester, N.Y., 1981.

Schechter, Harold, and David Everitt: *Film Tricks,* Quish Books, New York, 1981.

34 Photography: Still

Benedict, Joel A. and Douglas A. Crane: *Producing Multi-Image Presentations,* Arizona State University, Tempe, 1976.

Bullough, Robert V.: *Multi-Image Media,* Educational Technology Publications, Englewood Cliffs, N.J., 1981.

Cyr, Don: *Teaching Your Children Photography,* American Photography Book Publishing Company, Garden City, New York, 1980.

Pictures and Words, Centre for Advanced TV Studies, London, 1980.

Sunier, John: *Slide/Sound and Filmstrip Production,* Focal Press, New York, 1981.

35 Puppets and Puppetry

Jenkins, Peggy Davison: *Magic of Puppetry,* Prentice-Hall, New York, 1980.

Puppet Programs, kit with audio cassettes, Pretend Time Puppets, Boca Raton, Fla., 1981.

Puppets and You in the Library, Orange County Public Library, Orange, Calif., 1981.

36 Selection Aids: General

See the set of media reference books published and regularly revised by the National Center for Educational Media (NICEM), University of Southern California, University Park, Los Angeles, 90007:

Index to Educational Audio Tapes

Index to 16mm Educational Films

Index to Educational Overhead Transparencies

Index to Educational Resources

Index to Educational Slides

Index to Educational Video Tapes

Index to 8mm Motion Cartridges

Index to Environmental Studies—Multimedia

Index to Health and Safety Education—Multimedia

Index to Nonprint Special Educational Materials—Multimedia

Index to Producers and Distributors

Index to Psychology—Multimedia

Index to 35mm Filmstrips

Index to Vocational and Technical Education—Multimedia

Brown, James W., and Shirley N. Brown (eds.): *Educational Media Yearbook,* Libraries Unlimited Inc., Littleton, Col. A sourcebook of general information, including bibliographic and mediagraphic listings of items included in the Correlated References section of this manual.

Cabeceiras, James: *The Multimedia Library: Materials Selection and Use,* Academic Press, New York, 1978.

Selecting Media for Learning, Association for Educational Communications and Technology, Washington, D.C., 1974.

37 Sketching

Art Elements: An Introduction, 16mm film, 16 min., BFA Educational Media, Santa Monica, Calif., 1981.

Drawing People: An Introduction to Figure Drawing, 35mm sound filmstrips, Educational Dimensions Group, Stamford, Conn., 1981.

Frye, Harvey, and Ed Minor: *Techniques for Producing Visual Instructional Media,* McGraw-Hill, New York, 1977.

38 Slides: 2- by 2-inch

Beatty, LaMond F.: *Still Pictures,* Educational Technology Publications, Inc., Englewood Cliffs, N.J., 1981.

Kueter, Roger A.: *Slides,* Educational Technology Publications, Inc., Englewood Cliffs, N.J., 1981.

Saxby, Graham: *Focalguide to Slides,* Focal Press, New York, 1979.

39 Still Pictures: Flat, Opaque

Beatty, LaMond F.: *Still Pictures,* Educational Technology Publications, Englewood Cliffs, N.J., 1981.

Kemp, Jerrold E.: *Planning and Producing Audiovisual Materials,* Harper and Row, New York, 1980.

Williams, Catherine M.: *Learning from Pictures,* Association for Educational Communications and Technology, Washington, D.C., 1963.

40 Still Pictures: Laminating, Mounting

Kemp, Jerrold E.: *Planning and Producing Audiovisual Materials,* Harper and Row, New York, 1980.

Minor, Ed, and Harvey R. Frye: *Techniques for Planning and Producing Visual Instructional Materials,* New York: McGraw-Hill, 1970.

Morlan, John E.: *Preparation of Inexpensive Teaching Materials,* Crowell, New York, 1973.

41 Television

Block, David C.: "How to Plug Into Your Local Umbilical Cord: Everything You Need to Know About Cable Television," *Instructional Innovator,* February 1981, pp. 23–25ff.

Children and Television: The Development of a Child's Understanding of the Medium (ERIC ED 184 515), microfiche, ERIC, Washington, D.C., 1979.

Clement, Frank: "Oh Dad, Poor Dad, Mom's Bought the Wrong Videodisc and I'm Feelin' So Sad," *Instructional Innovator,* February 1981, pp. 12–15.

"Directory of Videodisc Players," *Educational and Industrial Television,* March 1981, pp. 40–53.

Hoglin, John Giles: "How to Use Low-Cost Video for Instruction," *Instructional Innovator,* January 1982, pp. 39–43.

Levinson, Richard, and William Link: *Stay Tuned: An Inside Look at the Making of Prime-Time Television,* St. Martin's Press, New York, 1981.

Littner, Ner: *Television Viewing and Its Effects—If Any—on Children,* Television Information Office, New York, 1980.

Pekich, Joe: *Instructional Television: Potentials or Problems?* Information Futures, Pullman, Wash., 1979.

Schramm, Wilbur, et al.: *Bold Experiment: The Story of Educational Television in American Samoa,* Stanford University Press, Stanford, 1981.

Sigel, Efrem, et al.: *Video Discs: The Technology, the Applications and the Future,* Knowledge Industry Publications, White Plains, N.Y., 1980.

Stier, Karen: "Video—From Ground Zero," *Audiovisual Directions,* pp. 28–31.

Thomas, Willard: "Interactive Video," *Instructional Innovator,* February 1981, pp. 19–20.

42 Transparencies: Large

Basic Tips on Producing and Using Overhead Transparencies, National Audiovisual Association, Fairfax, Va., 1981.

I Like the Overhead Projector Because..., Association for Educational Communications and Technology, Washington, D.C., 1977.

Kemp, Jerrold E.: *Planning and Producing Audiovisual Materials,* Row Peterson, New York, 1980.

Overhead Projection Guide, Arkwright Publications, Fiskerville, R.I., 1980.

Present Information with Overhead and Opaque Materials, Center for Vocational Education, American Association for Vocational Instructional Materials, Athens, Ga., 1977.

43 Visual Literacy

Debes, John: "Some Foundations for Visual Literacy," *Audiovisual Instruction,* Nov. 1978, pp. 961–964.

Dondis, Donis A.: *A Primer of Visual Literacy,* Massachusetts Institute of Technology, Cambridge, 1973.

Grammar of Media Kit, Hayden Book Co., Rochelle Park, N.J. 1980.

Sless, David, *Learning and Visual Communication,* Halsted Press/John Wiley and Sons, New York, 1981.

PERFORMANCE CHECKSHEETS

The following Performance Checksheets are designed to guide your practice with equipment, facilitate your learning recommended step-by-step procedures, and provide space for a record of your practice. The checksheets are to be used in ways most helpful to you as you practice with each machine. The times indicated are only suggestions, based upon performance of many students.

The Summary Performance checksheet is for a record of your final performance on each item of equipment studied, whether checkout is by an instructor, a laboratory assistant, a classmate, or yourself.

NOTE: References on the performance checksheets are coded as follows:

AVITM: Brown, J. W., and R. B. Lewis (eds.): *The AV Instructional Technology Manual* (6th ed.), McGraw-Hill, New York, 1983.

BLH: Brown, J. W., R. B. Lewis, and F. F. Harcleroad: *AV Instruction: Technology, Media, and Methods* (6th ed.), McGraw-Hill, New York, 1983.

FORMAT GUIDES

When you are preparing visuals for projection, guides that show the proportional dimensions required for each medium will save production time. Included here is an actual-size guide for overhead projector transparencies and guides in suitable proportions (height to width ratios) for slides, filmstrips, motion pictures, and television.

To use the format guides you may work directly over the guides in this manual, duplicate them on paper to place under your work, or duplicate them on clear plastic to improve light transmission through work done on a light table.

Remember to observe desirable lettering sizes and styles for legibility in each format. Remember also to leave adequate margins around your image to allow for cropping that might reduce the format size when the material is projected or reproduced.

AUDIO DISC PLAYERS

Name _____ Course _____ Date _____

References: *AVITM* Ex. 52
      BLH Ref. Sect. 1

TRIALS

	1	2	3	

Setting up (Time: 1 min)   **NOTES**
    Place and assemble ....
    Connect cords...................................
    Amplifier on ....
    Turntable motor on ....
                               Time

Operating (Time: 2 min)
    Select speed ....
    Check speed ....
    Select stylus ....
    Select arm weight ....
    Play stylus on record ....
    Play (1 minute).................................
    Adjust volume ....
    Adjust tone ....
    Lift arm ....
    Handle record ....
    Use microphone ....
    Amplify radio or tape recorder ....
                               Time

Putting away (Time: 1 min.)
    Lock arm ....
    Motor off ....
    Release turntable ....
    Amplifier off ....
    Store cord(s) ....
    Lock lid(s) ....
                               Time
                    TOTAL TIME

OPEN-REEL TAPE RECORDERS

Name _____ Course _____ Date _____

References: *AVITM* Ex. 53
BLH Ref. Sect. 1

	TRIALS			NOTES
	1	**2**	**3**	
Setting up (Time: 1 min)				
Place and assemble	____	____	____	
Amplifier and motor on	____	____	____	
Set speed control	____	____	____	
Check reel rotation	____	____	____	
Attach empty reel	____	____	____	
Attach supply reel	____	____	____	
Thread and attach tape	____	____	____	
Time	____	____	____	_____
				_____
Operating (Time: 1 min)				
Play	____	____	____	_____
Adjust volume and tone	____	____	____	_____
Rewind	____	____	____	_____
Time	____	____	____	_____
				_____
Recording (Time: 3 min)				
Connect microphone	____	____	____	_____
Test level	____	____	____	
Record (1 minute)	____	____	____	
Rewind, Play	____	____	____	
Stop, Rewind	____	____	____	
Time	____	____	____	_____
				_____
Putting away (Time: 1 min)				
Amplifier off	____	____	____	_____
Stow microphone	____	____	____	
Stow reels	____	____	____	
Stow cords	____	____	____	
Lock lid(s)	____	____	____	
Time	____	____	____	_____
TOTAL TIME	____	____	____	
Combination uses				
With another recorder	____	____	____	_____
With synch unit and projector	____	____	____	
With a record player	____	____	____	
With a radio	____	____	____	_____
				_____
Optionals				_____
Clean heads	____	____	____	
Splice tape	____	____	____	
Degauss heads	____	____	____	

PERFORMANCE CHECKSHEET 3

AUDIO CASSETTE TAPE RECORDERS

Name _____ Course _____ Date _____

References: *AVITM* Ex. 54
BLH Ref. Sect. 1

	TRIALS			
	1	**2**	**3**	**NOTES**
Setting up (Time: 1 min)				
Select power source	___	___	___	
Battery	___	___	___	
AC adapter or cord	___	___	___	
Time	___	___	___	_____
Operating (Time: 3 min)				_____
Play prerecorded cassette...........................	___	___	___	
Attach microphone..................................	___	___	___	
Record (1 minute)	___	___	___	
Rewind..	___	___	___	
Play	___	___	___	
Adjust volume and tone	___	___	___	
Use earphone	___	___	___	
Rewind..	___	___	___	
Time	___	___	___	_____
Putting away (Time: 1 min)				_____
Rewind fully	___	___	___	
Remove cassette	___	___	___	
Remove any accessories	___	___	___	
Close case and store parts	___	___	___	
Time	___	___	___	_____
TOTAL TIME	___	___	___	
Combination uses				_____
With another recorder	___	___	___	
With synch unit and projector	___	___	___	
With a record player	___	___	___	
With a radio	___	___	___	_____
Optionals				_____
Clean heads	___	___	___	
Degauss heads	___	___	___	
Change batteries	___	___	___	_____

OPAQUE PROJECTORS

Name _____ Course _____ Date _____

References: *AVITM* Ex. 60
BLH Ref. Sect. 1

	TRIALS			**NOTES**
	1	**2**	**3**	
Setting up (Time: 1 min)				
Connect power	_____	_____	_____	
Turn on lamp	_____	_____	_____	
Insert flat material	_____	_____	_____	
Elevate and level	_____	_____	_____	
Adjust focus	_____	_____	_____	
Time	_____	_____	_____	_____
Operating (Time: 3 min)				_____
Show flat pictorials (3)	_____	_____	_____	
Show book pages (3)	_____	_____	_____	
Show small objects	_____	_____	_____	
Use·pointer	_____	_____	_____	
Use conveyer belt	_____	_____	_____	
Time	_____	_____	_____	
Putting away (Time: 1 min)				_____
Retract lens	_____	_____	_____	
Store cord	_____	_____	_____	
Time	======	======	======	_____
TOTAL TIME	_____	_____	_____	
Optional				_____
Change lamp (when cool)	_____	_____	_____	

PERFORMANCE CHECKSHEET 5

OVERHEAD PROJECTORS

Name _____ **Course** _____ **Date** _____

References: *AVITM* Ex. 56
BLH Ref. Sect. 1

TRIALS

	1	2	3	NOTES
(Time: 2 min*)				
Set up and assemble	_____	_____	_____	
Put on test transparency	_____	_____	_____	
Plug in and turn on	_____	_____	_____	
Adjust top mirror	_____	_____	_____	
Adjust image size	_____	_____	_____	
Focus	_____	_____	_____	
Replace test with demonstration transparency	_____	_____	_____	
Manipulate pointer, overlays, masks	_____	_____	_____	
Shut down and put away	_____	_____	_____	
Time	=====	=====	=====	_____
TOTAL TIME	_____	_____	_____	

Optional _____
  Change lamp (when cool) . . . . _____ _____ _____
* Assumes screen is in place.

2- BY 2-INCH SLIDE PROJECTORS

Name _____ Course _____ Date _____

References: *AVITM* Ex. 57
BLH Ref. Sect. 1

	TRIALS			
	1	**2**	**3**	**NOTES**

Setting up (Time: 2 min)
Load slide carrier . . . .
Place projector . . . .
Connect power cord . . . .
Connect remote cord . . . .
Install slide carrier . . . .
Center light on screen . . . .
Adjust image size . . . .
Focus . . . .
Time

Operating (Time: 2 min)
Turn on fan and lamp . . . .
Advance slides . . . .
Refocus as necessary . . . .
Skip slides . . . .
Go back two slides . . . .
End showing . . . .
Time

Putting away (Time: 1 min)
Disconnect cords . . . .
Stow cords . . . .
Remove and stow carrier . . . .
Close case and latch lids . . . .
Time
TOTAL TIME

Combination uses
Connect and use with slide synchronizer . . . .

Optionals
Replace lamp (when cool) . . . .
Clean lens . . . .

FILMSTRIP PROJECTORS

Name _____ Course _____ Date _____

References: *AVITM* Ex. 59
BLH Ref. Sect. 1

	TRIALS			NOTES
	1	2	3	
Setting up (Time: 1 min)				
Place and assemble	___	___	___	
Center light on screen	___	___	___	
Focus light beam	___	___	___	
Time	___	___	___	_____
Operating (Time: 3 min)				_____
Handling filmstrip	___	___	___	
Insert filmstrip	___	___	___	
Correct focus	___	___	___	
Project filmstrip	___	___	___	
Adjust frame line	___	___	___	
Back up two frames	___	___	___	
Skip two frames	___	___	___	
Close showing	___	___	___	
Rewind filmstrip	___	___	___	
Time	___	___	___	_____
Changing to 2- by 2-inch slides (Time: 2 min)				
Remove filmstrip carrier	___	___	___	
Insert slide carrier	___	___	___	
Project for image size	___	___	___	
Focus	___	___	___	
Project slides (5)	___	___	___	
Handle slides	___	___	___	
Time	___	___	___	_____
Putting away (Time: 1 min)				_____
Disassemble	___	___	___	
Stow cord	___	___	___	
Stow accessories	___	___	___	
Latch lid	___	___	___	
Stow all slides and strips	___	___	___	
Time	___	___	___	_____
TOTAL TIME	___	___	___	
Optionals				_____
Clean film channel	___	___	___	
Clean lens	___	___	___	
Change lamp (when cool)	___	___	___	_____

16MM SOUND PROJECTORS, MANUAL-THREADING

Name _____ Course _____ Date _____

References: *AVITM* Ex. 61, 62
BLH Ref. Sect. 1

	TRIALS			NOTES
	1	**2**	**3**	
Setting up (Time: 3 min)				
Place on stand	⎯⎯	⎯⎯	⎯⎯	
Open case	⎯⎯	⎯⎯	⎯⎯	
Connect cords.....................................	⎯⎯	⎯⎯	⎯⎯	
Amplifier on	⎯⎯	⎯⎯	⎯⎯	
Attach or set arms	⎯⎯	⎯⎯	⎯⎯	
Check speed and rotation	⎯⎯	⎯⎯	⎯⎯	
Lamp on, center on screen	⎯⎯	⎯⎯	⎯⎯	
Focus aperture image	⎯⎯	⎯⎯	⎯⎯	
Clean gate	⎯⎯	⎯⎯	⎯⎯	
Attach reels	⎯⎯	⎯⎯	⎯⎯	
Thread film	⎯⎯	⎯⎯	⎯⎯	
Recheck threading	⎯⎯	⎯⎯	⎯⎯	
Time	⎯⎯	⎯⎯	⎯⎯	_____
Operating (Time: 15 sec*)				_____
Preset volume and tone	⎯⎯	⎯⎯	⎯⎯	
Motor on, lamp on................................	⎯⎯	⎯⎯	⎯⎯	
Focus	⎯⎯	⎯⎯	⎯⎯	
Adjust volume and tone	⎯⎯	⎯⎯	⎯⎯	
Adjust frame line	⎯⎯	⎯⎯	⎯⎯	
Closing showing (as appropriate):				
Sound down, lamp off	⎯⎯	⎯⎯	⎯⎯	
Motor off	⎯⎯	⎯⎯	⎯⎯	
Rewind film	⎯⎯	⎯⎯	⎯⎯	
Time	⎯⎯	⎯⎯	⎯⎯	_____
Putting away (Time: 2 min)				_____
Disconnect cords	⎯⎯	⎯⎯	⎯⎯	
Disassemble equipment	⎯⎯	⎯⎯	⎯⎯	
Stow parts	⎯⎯	⎯⎯	⎯⎯	
Retract elevating device	⎯⎯	⎯⎯	⎯⎯	
Lock lids	⎯⎯	⎯⎯	⎯⎯	
Time	⎯⎯	⎯⎯	⎯⎯	_____
TOTAL TIME	⎯⎯	⎯⎯	⎯⎯	

Optionals

Handle film	⎯⎯	⎯⎯	⎯⎯	
Change projection lamp (when cool)	⎯⎯	⎯⎯	⎯⎯	
Change exciter lamp	⎯⎯	⎯⎯	⎯⎯	
Clean lens	⎯⎯	⎯⎯	⎯⎯	_____

* Not including short film sequence shown.

16MM SOUND PROJECTORS, AUTOMATIC AND SLOT-THREADING

Name _____ Course _____ Date _____

References: *AVITM* Ex. 61, 62
BLH Ref. Sect. 1

	TRIALS			NOTES
	1	2	3	
Setting up (Time: 2½ min)				
Place on stand	____	____	____	
Open case	____	____	____	
Connect cords	____	____	____	
Amplifier on	____	____	____	
Attach or set arms	____	____	____	
Check speed and rotation	____	____	____	
Lamp on, center on screen	____	____	____	
Focus aperture image	____	____	____	
Clean gate	____	____	____	
Attach reels	____	____	____	
Set auto threading	____	____	____	
Trim leader	____	____	____	
Projector on	____	____	____	
Thread film	____	____	____	
Attach to take-up reel	____	____	____	
Time	____	____	____	_____
Operating (Time: 15 sec*)				_____
Preset volume and tone	____	____	____	
Motor on, lamp on	____	____	____	
Focus	____	____	____	
Adjust volume and tone	____	____	____	
Adjust frame line	____	____	____	
Closing showing (as appropriate):	____	____	____	
Sound down, lamp off	____	____	____	
Motor off	____	____	____	
Rewind film	____	____	____	
Time	____	____	____	_____
Putting away (Time: 2 min)				_____
Disconnect cords	____	____	____	
Disassemble equipment	____	____	____	
Stow parts	____	____	____	
Retract elevating device	____	____	____	
Lock lids	____	____	____	
Time	____	____	____	_____
TOTAL TIME	____	____	____	
Optionals				_____
Handle film	____	____	____	
Change projection lamp (when cool)	____	____	____	
Change exciter lamp	____	____	____	
Clean lens	____	____	____	

* Not including short film sequence shown.

PERFORMANCE CHECKSHEET 10

TELEVISION RECEIVERS

Name _____ Course _____ Date _____

References: *AVITM* Ex. 64
BLH Ref. Sect. 1

<table>
<tr><td></td><td colspan="3" align="center">TRIALS</td><td></td></tr>
<tr><td></td><td>1</td><td>2</td><td>3</td><td></td></tr>
<tr><td>Operating (Time: 2 min)</td><td></td><td></td><td></td><td align="right">NOTES</td></tr>
<tr><td>Turn on.............................</td><td>____</td><td>____</td><td>____</td><td></td></tr>
<tr><td>Select channel ....</td><td>____</td><td>____</td><td>____</td><td></td></tr>
<tr><td>Adjust fine tuning ....</td><td>____</td><td>____</td><td>____</td><td></td></tr>
<tr><td>Adjust brightness ....</td><td>____</td><td>____</td><td>____</td><td></td></tr>
<tr><td>Adjust contrast ....</td><td>____</td><td>____</td><td>____</td><td></td></tr>
<tr><td>Adjust vertical...................</td><td>____</td><td>____</td><td>____</td><td></td></tr>
<tr><td>Correct horizontal...............</td><td>____</td><td>____</td><td>____</td><td></td></tr>
<tr><td>Adjust tint ....</td><td>____</td><td>____</td><td>____</td><td></td></tr>
<tr><td>Adjust color and contrast ....</td><td>____</td><td>____</td><td>____</td><td></td></tr>
<tr><td>Adjust volume and tone ....</td><td>____</td><td>____</td><td>____</td><td></td></tr>
<tr><td>Shut off ....</td><td>____</td><td>____</td><td>____</td><td></td></tr>
<tr><td align="right">Time</td><td>____</td><td>____</td><td>____</td><td>_____</td></tr>
</table>

TELEVISION CAMERAS

Name _____ Course _____ Date _____

References: *AVITM* Ex. 66
BLH Ref. Sect. 1

	TRIALS			
	1	**2**	**3**	
Setting up (Time*)				**NOTES**
Camera on tripod or stand	____	____	____	
Set lights in position	____	____	____	
Connect and turn on:				
Camera	____	____	____	
Receiver	____	____	____	
Time	____	____	____	_____
Operating (Time*)				_____
Place object in field	____	____	____	
Adjust lights	____	____	____	
Adjust				
Focus	____	____	____	
Brightness	____	____	____	
Contrast	____	____	____	
Demonstrate showing object	____	____	____	
Demonstrate showing flat picture or drawing	____	____	____	
Time	____	____	____	_____
Shutting down (Time*)				_____
Turn off all components	____	____	____	
Cover lens	____	____	____	
Disassemble as required	____	____	____	
Stow all parts	____	____	____	
Time	____	____	____	_____
TOTAL TIME	____	____	____	

* Variations in equipment and conditions of practice markedly affect performance time. Determine reasonable time limits locally.

VIDEO TAPE RECORDING SYSTEMS

Name _____ Course _____ Date _____

References: *AVITM* Ex. 66
BLH Ref. Sect. 1

	TRIALS			**NOTES**
	1	2	3	
Setting up (Time*)				
Set up components	____	____	____	
Interconnect units	____	____	____	
Handle tape.......................................	____	____	____	
Thread tape	____	____	____	
Connect and turn on power	____	____	____	
Adjust camera for focus	____	____	____	
Adjust brightness and contrast	____	____	____	
Check audio	____	____	____	
Time	____	____	____	_____
Operating (Time*)				_____
Record (2 minutes)	____	____	____	
Stop and Rewind.................................	____	____	____	
Play and Evaluate	____	____	____	
Recorder controls	____	____	____	
Camera manipulation	____	____	____	
Picture quality...............................	____	____	____	
Time	____	____	____	_____
Putting away (Time*)				_____
Turn off all switches	____	____	____	
Stow all parts properly	____	____	____	
Latch cases firmly	____	____	____	
Time	____	____	____	_____
TOTAL TIME	____	____	____	
Optionals				
Clean heads	____	____	____	
Clean camera lens	____	____	____	_____

* Variations in equipment and conditions of practice markedly affect performance time. Determine reasonable time limits locally.

MICROCOMPUTERS

Name _____ Course _____ Date _____

References: *AVITM* Ex. 53
BLH Ref. Sect. 1

	TRIALS			
	1	2	3	**NOTES**

Preliminary
Arrange components . . . .
Be sure On-Off switch is off . . . .
Plug in AC line cord . . . .
Remove cover from computer . . . .
Touch metal case of power supply (to drain off any body static) . . . .

Operating (Preliminary)
Install disk, if used . . . .
Install drive controller card (in the microcomputer "mother board") . . . .
Attach the cable for CRT monitor . . . .
Attach printer cable to printer interface card . . . .
Attach printer cable to printer . . . .
Insert printer interface card in slot provided on "mother board" . . . .
Replace cover on microcomputer . . . .

Operating
Turn on CRT monitor . . . .
Insert program disk into disk drive; close door . . . .
Turn on microcomputer . . . .
Adjust CRT monitor (brightness and contrast) for best resolution . . . .
Turn on printer . . . .

Closing down
Remove disk from disk drive . . . .
Turn off CRT monitor . . . .
Turn off printer . . . .
Turn off microcomputer . . . .

SPIRIT DUPLICATORS—MASTERS

Name _____ Course _____ Date _____

References: *AVITM* Ex. 33
BLH Ref. Sect. 1

TRIALS

	1	2	3	NOTES

Prepare the master pack
Obtain master paper-carbon set ....
Remove protective slip sheet ....
Attach backing sheet ....

Type masters
Insert pack into typewriter.........................
Type with purple carbon ....
Type with other colors ....
Type justified copy specimen ....

Make corrections
Correct by strikeovers ....
Correct by cutouts ....
Correct by tape overlays ....
Correct by scraping ....
Correct by wax pencil...............................
Correct with new carbon ....

Thermal masters
Select master assembly ....
Set copier temperature (time) ....
Arrange, run thermal master ....
Peel master; check quality ....

Hand operations
Trace line illustrations on master ....
Do freehand drawing ....
Do hand lettering ....
Use lettering guide ....

Cutting, reassembling
Cut master; eliminate section ....
Reassemble ....
Cut master; eliminate section ....
Reassemble ....
Overlay entire section with tape ....

SPIRIT DUPLICATORS

Name _____ Course _____ Date _____

References: *AVITM* Ex. 33
BLH Ref. Sect. 2

	TRIALS			NOTES
	1	2	3	

Ready the duplicator
Remove dust cover . . . . _____ _____ _____
Clean duplicator if required . . . . _____ _____ _____
Insert paper in feed tray . . . . _____ _____ _____
Adjust paper guides, grippers . . . . _____ _____ _____
Adjust pressure lever . . . . _____ _____ _____
Check wick for moistness . . . . _____ _____ _____ _____

Attach master
Turn drum to proper position . . . . _____ _____ _____
Depress master clamp lever . . . . _____ _____ _____
Insert master . . . . _____ _____ _____
Return clamp to locked position . . . . _____ _____ _____ _____

Run copies
Run a test sheet . . . . _____ _____ _____
Adjust margins (top, bottom, sides) . . . . _____ _____ _____
Adjust impression strength . . . . _____ _____ _____
Check spirit flow; adjust if needed . . . . _____ _____ _____
Set counter for copies desired . . . . _____ _____ _____
Run copies; check continuously . . . . _____ _____ _____ _____

Leaving the machine
Turn drum to proper position . . . . _____ _____ _____
Open clamp lever; remove master . . . . _____ _____ _____
Cover master with specimen sheet; file . . . . _____ _____ _____
Close clamp lever . . . . _____ _____ _____
Set copy control lever to 0 . . . . _____ _____ _____
Set pressure control lever to 0 . . . . _____ _____ _____
Remove unused paper . . . . _____ _____ _____
Remove and store spirit reservoir . . . . _____ _____ _____
Leave drum in proper position . . . . _____ _____ _____
Replace dust cover . . . . _____ _____ _____ _____

SUMMARY

Name _____ Course _____ Date _____

Equipment items	PRACTICE TRIALS COMPLETED			FINAL DATE	TRIAL SCORE
1 Record players	_____	_____	_____	_____	_____
2 Open Reel Tape Recorders	_____	_____	_____	_____	_____
3 Audio Cassette Tape Recorders	_____	_____	_____	_____	_____
4 Opaque Projectors	_____	_____	_____	_____	_____
5 Overhead Projectors	_____	_____	_____	_____	_____
6 2- by 2-Inch Slide Projectors	_____	_____	_____	_____	_____
7 Filmstrip Projectors	_____	_____	_____	_____	_____
8 16mm Sound Projectors, Manual	_____	_____	_____	_____	_____
9 16mm Sound Projectors, Automatic and Slot-Threading	_____	_____	_____	_____	_____
10 Television Receivers	_____	_____	_____	_____	_____
11 Television Cameras	_____	_____	_____	_____	_____
12 Video Tape Recording Systems	_____	_____	_____	_____	_____
13 Microcomputers	_____	_____	_____	_____	_____
14 Spirit Duplicators—Masters	_____	_____	_____	_____	_____
15 Spirit Duplicators	_____	_____	_____	_____	_____

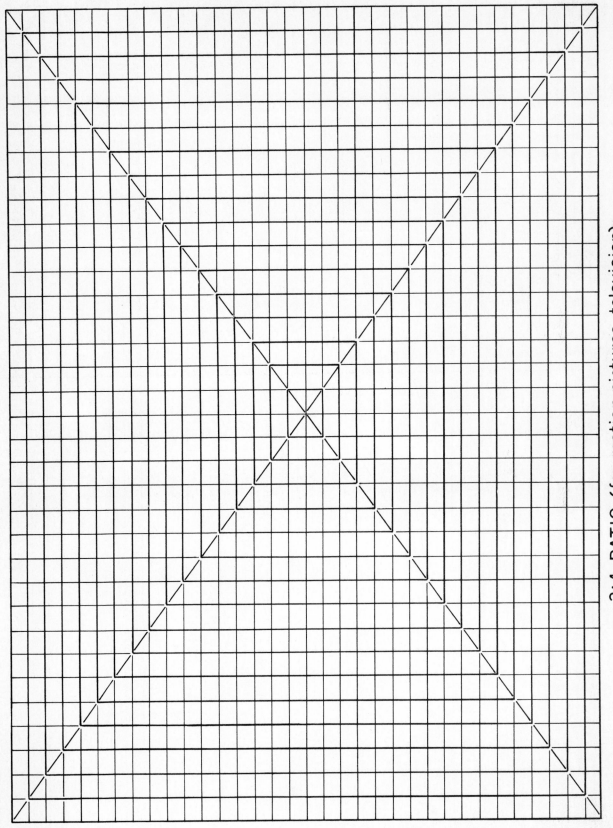

3:4 RATIO (for motion pictures, television)

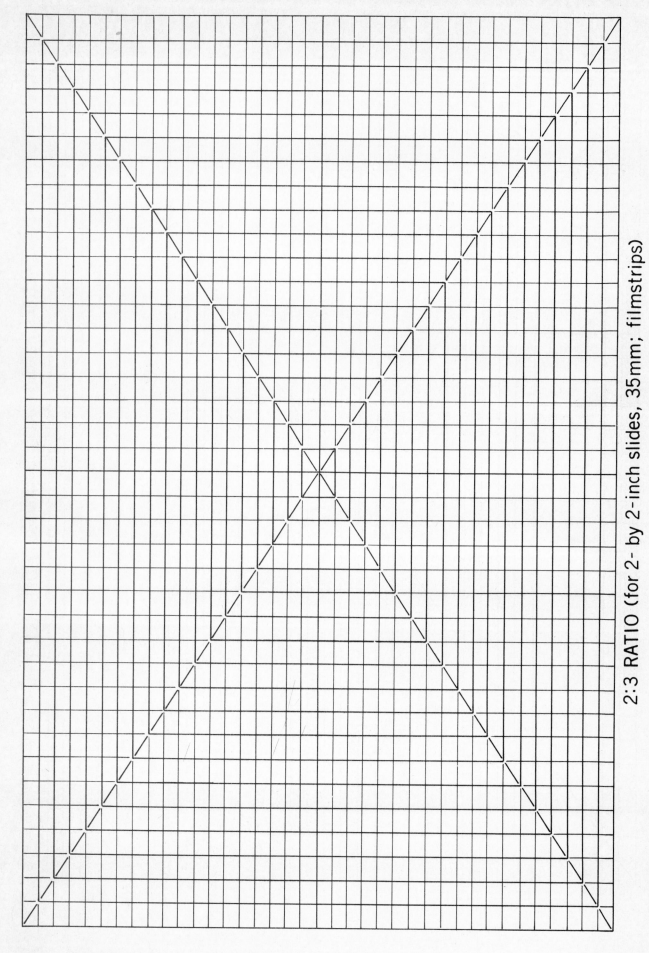

2:3 RATIO (for 2- by 2-inch slides, 35mm; filmstrips)